Pass That Math Test

Our Additional Evidence-Based Resource Books for Educators

Handbook on Effective Instructional Strategies: Evidence for Decision-Making
Myles I. Friedman and Steven P. Fisher

Ensuring Student Success: A Handbook of Evidence-Based Strategies
Myles I. Friedman

Educators' Handbook on Effective Testing
Myles I. Friedman, Charles W. Hatch, Jacqueline E. Jacobs, Aileen C. Lau-Dickinson, Amanda B. Nickerson, and Katherine C. Schnepel

No School Left Behind: How to Increase Student Achievement
Myles I. Friedman

Effective Instruction: A Handbook of Evidence-Based Strategies
Myles I. Friedman, Diane H. Harwell, and Katherine C. Schnepel

Developing Teaching Effectiveness
Myles I. Friedman, Diane H. Harwell, and Katherine C. Schnepel

Developing Teaching Effectiveness: Instructor's Manual
Myles I. Friedman, Diane H. Harwell, and Katherine C. Schnepel

Pass That Test: A Guide to Successful Test Taking
Charles W. Hatch

Pass That Math Test

Success with PRAXIS, SAT, ACT, GRE, GMAT, GED, and State Exit Examinations

Charles W. Hatch and Micki Durham Gibson

THE INSTITUTE FOR EVIDENCE-BASED
DECISION-MAKING IN EDUCATION, INC.

Copyright © 2011 by
The Institute for Evidence-Based Decision-Making in Education, Inc.

All rights reserved. No portion of this book may be
reproduced, by any process or technique, without the
express written consent of the publisher.

Library of Congress Control Number: 2010943470
ISBN: 978-0-9666588-8-0

First published in 2011

The Institute for Evidence-Based Decision-Making in Education, Inc.
A South Carolina non-profit corporation
P.O. Box 122, Columbia, SC 29202

Printed in the United States of America

The paper used in this book complies with the
Permanent Paper Standard issued by the National
Information Standards Organization (Z39.48–1984).

10   9   8   7   6   5   4   3   2   1

Dedicated to those who have inspired us.
C. W. H. and M. D. G.

Contents

Preface		ix
1.	Test Constructors' Assumptions	1
2.	Basics of Multiple-Choice Questions	6
3.	The Nature of Numbers	16
4.	Number Line/Place Value	22
5.	Decimals	32
6.	Rounding	35
7.	Signs	38
8.	Fundamental Operations: Addition and Subtraction	49
9.	Fundamental Operations: Multiplication and Division	53
10.	Fractions: Addition and Subtraction	64
11.	Fractions: Multiplication and Division	68
12.	Fractions to Decimals to Percents	80
13.	Ratios and Proportions	94
14.	Statistics	100
15.	Geometry, Basic Angles, and Figures	110
16.	Geometry: Perimeter	149
17.	Geometry: Area	152
18.	The Pythagorean Theorem	158

19.	Circles	168
20.	Probability	173
21.	Decoding Tables and Graphs	179
22.	Distance, Rate, and Time Problems	207
23.	Work Problems	215
24.	Elementary Algebra	219
25.	Roots, Powers, and Scientific Notation	239
26.	Inequalities	247
27.	Age Problems	255
28.	Trigonometry	260
29.	Measurement	268

About the Authors   271

Preface

The authors of this book had two keys in mind as they wrote.

FIRST KEY

The first key is attitudinal and abstract. We feel that passing these tests is *not* related to the amount of difficult math that you can master. Instead, passing is dependent on your getting correct nearly all of the questions related to math concepts you already know and understand. Many test takers miss question after question on material they actually grasp well. For example, one person might miss a question through carelessness with signs; another might miss one of those d = rt questions such as the one featured in Chapter 1. All of these are points the test taker deserves but does not get.

This book's purpose is to review, explore, and present fundamental concepts to raise test scores.

SECOND KEY

The second key is purely practical. We recommend that you study this book with a large 4 × 6 index card or sheet of paper to slide down each page to cover the answers. We want to encourage you to work out each question without looking at our answer. This is certainly harder. We realize that, but the results will be significantly better. You will learn much less by just reading this book.

NOTATION

We have used the two symbols for multiplication (× and •) interchangeably. We have also used the two symbols for division (÷ and /) in the same manner.

1

Test Constructors' Assumptions

The people who design standardized tests make a number of assumptions about the individuals taking each test. These points make a useful initial checklist for test takers.

First, they assume that each testee has absorbed the free test descriptors. For example, PRAXIS makes available "Test at a Glance" (TaaG) descriptors, which can be downloaded from the PRAXIS Web site, usually accessed through the Educational Testing Service (ETS) homepage at www.ets.org. Each TaaG has a wonderfully helpful section titled "Topics Covered," which provides detailed information on the test content. A test taker preparing to take the PRAXIS who has not spent considerable time with the appropriate TaaG will not be fully prepared. A number of sample questions are included in each TaaG. Test takers are advised only to glance over these and to spend little time studying them.

Second, they assume that appropriate courses have been taken that cover most if not all of the test topics. However, this may not be true of all test takers, in which case the burden falls on those persons to learn those areas of weakness. This is especially noticeable in persons who are attempting a parallel employment move, such as from engineering to teaching high school math. A related factor is the quality of instruction that a given test taker has had. Because PRAXIS I draws largely from math instruction given in grades 6–9, a deficient program at this level will handicap its students.

In a similar vein, all high school and post–high school programs are not equal. If a program is not rigorous in insisting that students understand material, SAT, ACT (American College Testing Assessment), and GRE (Graduate Record Exam) test results will be disappointing because each question assumes that a testee understands relevant concepts.

Third, the test makers assume that a testee has mastered the basic material starting with the math facts. Over the last few decades many students have not been able to satisfy even this basic requirement. When a student's fingers start jumping during a calculation, the prognosis is poor because that person is counting to get an answer and has not even mastered the facts. The ability to do long division is another basic skill often lacking in weaker test takers. The quickest way to find out if a given student has mastered basic math facts is to simply ask, "What is the product of seven and eight?" ($7 \times 8 = ?$). If a person knows this, the basic facts have probably been mastered; if not, you have a problem needing remediation.

Fourth, test takers must realize that the tests may not cover the content of the grades referenced in the test name. Exceptions to this are the PRAXIS Elementary Education Content Knowledge 0014 and the Middle Grades Content Knowledge 0146, which do actually cover the grades referenced.

An example will illustrate. The Middle Grades Math Test (0069) does not simply cover the math taught in grades 5 through 9. It actually covers somewhat higher levels in greater depth than is taught in most classes through high school. The same is true of the secondary math tests. The result is that it is not really sufficient to "study" the textbooks for the grades covered in the test name. The following table illustrates the approximate content breadth for each test.

Test Name	Test Number(s)	Covers Through
PRAXIS I Math	5730, 0730	About ninth grade
Elementary Education Content Knowledge	0014*	About sixth grade
Middle School Content Knowledge	0146*	About eighth grade
SAT/ACT		About twelfth grade
GED		About high school
GRE		About college junior
Math Content Knowledge	0061	About college sophomore
Math Proofs, etc.	0063	About college sophomore
Math Pedagogy	0065	About college sophomore
General Math	0067	About college sophomore

* *Note*: 0014 and 0146 are included because although they are not exclusively math tests, each contains a whole math calculation section that counts for one-fourth of the test questions.

Fifth, test makers assume that testees are able to respond to questions quickly and accurately. Since these are all timed tests, the test taker must be able to produce correct answers quickly. A good example of this can be gained by looking at the 0069 Middle School Math Test. In 120 minutes (2 hours) the following must be done: answer forty multiple-choice questions, some of which require considerable calculation, plus three constructed response questions requiring a number of component answers. For example, one part of one of these latter questions may require the graphing of a complex relationship. This level of complexity, coupled with the time restraints, may be beyond the skills of many test takers. It almost seems that the better prepared a person is for the math tests beyond PRAXIS I, the harder the tests are to complete. This is because it takes longer to answer a question where the processes are understood than it does to answer a question where a person has little or no understanding of the processes and the only possible response is a feeble guess.

The following typical example illustrates many of the necessary steps that a test taker must traverse to answer a question accurately.

Question: James and Robin begin driving in opposite directions on I-95 from Florence, South Carolina, one going north, one south. If James drives 60 miles per hour and Robin drives 70 miles per hour, when will they be 325 miles apart?

Answer: (1) The situation must be visualized like this.

(2) The d = rt formula must be *modified* to reflect the *two* drivers:

d = Robin's rate (70 mph) and time + James's rate (60 mph) and time

$d = r_r t_r + r_j t_j$.

325 = 70t + 60t (Total distance is given as are the two speeds, and since both will drive the same amount of time there is only one unknown—the time they drive.)

325 = 130t (In each hour they are 130 miles farther apart.)

325/130 = t (How many 130's in the total mileage?)

t = 2.5 (or 2 hours and 30 minutes to be 325 miles apart).

There is another, more intuitive approach to a problem like this that avoids the use of a formula but still gets the correct answer:

Step 1. Think that for each hour the two drivers are on the road, they are 130 (60 + 70) miles farther apart.

Step 2. How many 130-mile *units* are in the total distance? This will be the correct answer in hours.

325/130 = 2.5 (or, alternatively, 2 hours and 30 minutes)

Sixth, a test maker assumes that the test takers know the approximate level of mathematics knowledge required for the various math tests. This awareness allows a testee to study effectively for a satisfactory score. The figure below illustrates that information.

Relative Levels of Mathematics Required for Various Tests

Most Advanced
- 0061 Mathematics: Content Knowledge ⎫
- 0063 Mathematics Proofs, Models, . . . ⎬ about equal
- 0065 Mathematics Pedagogy ⎭
- GRE
- SAT/ACT
- GED
- 0067 General Mathematics
- 0069 Middle School Mathematics

- 0511 Fundamental Subjects: Content Knowledge
- 0146 Middle School Content Knowledge
- 0730, 5730 PRAXIS I Math

Most Elementary
- 0014 Elementary Education: Content Knowledge

Thus, the tests requiring the most basic familiarity with math are PRAXIS I and Elementary Education: Content Knowledge (0014). Scarcely to be distinguished but slightly higher is the Middle School Content Knowledge (0146). Next in complexity comes the Fundamental Subjects: Content Knowledge (0511), which, like the 0014 and 0146 tests, contains one section on math that comprises one-fourth of the test questions. The Middle School Math test (0069) requires a considerably higher level of math than any of the others. After that come the General Math test (0067) and the GED. Next are the SAT/ACT and the GRE. At the highest level are the three secondary math tests, in no particular order of difficulty: Mathematics: Content Knowledge (0061), Mathematics: Proofs, Models and Problems Part I (0063), and finally Mathematics Pedagogy (0065).

The Teaching Foundations: Mathematics (0068) test of the PRAXIS is not included in the figure above because it is fundamentally different from all the other tests in that it does not include any mathematical calculations. It only asks a test taker to construct two long essays (over one hour each) on the teaching methods at middle/junior high and high school levels. Each of the answers requires the inclusion of the following:

- Instructional sequence
- Participatory or group activities
- Activities to strengthen reading skills
- Assessment

Thus, there are a number of relevant assumptions made by the test constructors that must be taken into consideration in test preparation. They *cannot be ignored* because of their basic nature.

2

Basics of Multiple-Choice Questions

ABILITY TO "PICTURE" A TEST

A student came to us a short time ago for help with the PRAXIS tests. He had an uncommon problem—he could not picture what was wanted. In fact, on one of the tests he became so confused that he cancelled his scores!

We started with the Tests at a Glance (TaaGs). These publications contain information on the number of questions, content, and time constraints, as well as some sample questions. The key to succeeding at many of these tests is proper time management. Math tests in particular are subject to time management problems. It seems that the more familiar you are with the material, the harder time you have finishing. If you know the procedures involved in answering each question, applying this information takes time—more time, in fact, than simply guessing at an answer and moving on.

In our student's case, he was so thoroughly confused by the proctor's presentation that he could make no reasonable response. After an hour or so going over the formats, he was much more relaxed and knew how to visualize the tests and what he needed to do to be successful. Academically over the years, he had had to put forth more effort than his classmates, but he was willing to do that in order to become the first member of his family to graduate from college. From there he hoped to realize his dream of becoming a football coach.

GENERAL ADVICE TO TEST TAKERS

Many students need to be reminded that there are some ways to approach a question that will tend to raise scores. Often a gentle reminder will be enough to get them back on track with improved test-taking strategies.

First, although it seems obvious, they need to be reminded to answer all of the test questions. The only exception to that rule is in the case of multiple-choice tests that use a correction for guessing. In this case, when a student has no idea of the answer, leaving it blank is a viable alternative that will not lower a score. A correction for guessing simply means that the testing company subtracts a percentage of the incorrect answers from the correct ones.

Not leaving an answer blank also applies to discussion questions. There is no possibility of credit when there is no answer, so even a weak answer is preferable to none.

Second, we urge our students *not* to change multiple-choice answers. Even though we get some students whose second choice tends to be more accurate than the first, most people give their best choice first. We are always dismayed to ask a group about changing answers and find that they "know" they shouldn't change answers but do it anyway. Sometimes they make two or three changes on the same question! Beside the fact that most people are more likely to choose the correct answer first, an optical test scanner can be confused when it senses incomplete erasures and gives no credit. We often have students who fail PRAXIS by only a few points and change answers quite often. We can see how changing only a couple of answers from correct to incorrect could make all the difference between failing and passing.

A corollary to not changing answers is our advice *not* to go back over a multiple-choice test. The prime motivation in reviewing answers is a readiness to change. Changing tends to have negative consequences, so why go back? Where there is only a downside to an activity, why pursue it?

Third, we stress neatness and legibility. Essay answers need to be neat and readable, with few erasures. A short while ago we were working with a student in Mississippi preparing for the PRAXIS PLT, which contains twelve essay answers. The student gave us a sample answer to evaluate. Its appearance seemed fine but was absolutely indecipherable! The student could not even read it herself. This principle would especially apply to the constructed response questions on Middle School Math (0069), Math Pedagogy (0065), and Math Proofs, Models and Problems (0063).

Another aspect of neatness arises with multiple-choice answers where marks are made weakly or outside the indicated answer spots. For example, we have had a number of students who circled the indicated response instead of darkening it in. In this case, an optical scanner would be completely misled and the score lowered to random answers because the circles tended to cross answer spaces above and below where intended.

Related to this are multiple-choice questions where more than one response is given—multiple answers. Whenever the scanner senses this, no credit is given. Sometimes students change an answer and simply forget to erase the first choice. Sometimes they slip an answer onto the wrong line, inadvertently giving two answers. A third type is the most common. Here a student reads the question, decides on an answer, starts to mark the choice, changes his or her mind, and indicates another choice. The answer sheet then shows a small mark for the initial choice and a larger, darker mark for the second. Usually the first, smaller answer is the correct answer but the students have "talked" themselves out of the first answer. Probably between 5 and 10 percent of test takers lower their scores by committing one of the above blunders while taking a standardized test. Students can easily avoid lowering their scores through greater attention to these simple details.

SEEING THE STRUCTURES OF A MULTIPLE-CHOICE QUESTION

Virtually all multiple-choice questions have a few simple formats. Not all students are familiar and comfortable with all of them. It is even possible to draw helpful diagrams of questions during a test in order to arrive at an answer. This analytical procedure is of no help if the testee knows nothing about the content of a given question, but it will provide considerable assistance where there is some knowledge, even if incomplete. Since one of the main goals in successful test preparation has to be to have the test takers understand the question better, this technique will be of considerable importance. One of the constant refrains that we hear in letters from students who have improved their test performance is that they have "finally learned how to read a test question."

Structural Type I-A

This type of multiple-choice question is the least complex structurally and is also the type of question most often written by teachers when creating test questions. This structure has one correct response and three or four incorrect ones depending on the number of alternatives. The typical diagram looks like this:

Correct	Incorrect
―	―
	―
	―
	―

The following example will serve to illustrate.

Question: The square root of 64 is:

A. 7
B. 8
C. 32
D. 16
E. 6

Correct	Incorrect
B	A
	C
	D
	E

Remember, just because I-A questions are not structurally complex, it does not mean that they are low in difficulty. A common mistake that we see is that test takers think all multiple-choice questions are in this format and that their only task for a whole test is to select the single correct answer in each instance. Sadly, this is not true.

A variant of this type is the multi-level question, which may look something like this:

Prompt

I. Details

II. Details

III. Details

Possible answer choices are:

A. I

B. II

C. III

D. I and II

E. II and III

There is *one* correct answer, but the whole process is complicated by an additional step where you must decide on the Roman numeral responses before attempting the actual answers.

When we start talking about these multi-level questions in class, we invariably hear many students groan that they just hate them. This means that they approach them negatively, with no clear strategy. It also means that they miss a disproportionate number of them simply because of the complex format.

We advise "solving" the Roman numeral segment first, then moving to the choices for an answer. We even go so far as to cover the final choices with a hand so that we have to focus on the Roman numbered choices first. It helps us to mark each one of them as "T," "F," or "?" PRAXIS math tests (0061, 0069, etc.) contain these multi-level questions (see TaaGs). It is possible to answer these even *without* complete knowledge of each Roman-numbered choice. For example, if you could only decide that one choice is false, look down the choices and eliminate all those that contain that option. You may even surprise yourself and be able to narrow the posibilities to one! Processes of logic are important here.

Structural Type I-B

This type of question is the exact opposite of the I-A. That is, the testee's task is to find the single *incorrect* answer. In diagram form, the structure looks like this:

Correct	Incorrect
___	___
___	
___	
___	

The following worked example illustrates this:

Question: If there are exactly 5 times as many children as adults at a show, all of the following could be the number of people in the show EXCEPT:

A. 102
B. 80
C. 36
D. 30
E. 72

Because all of the above are divisible by six except B, that is the "incorrect" answer.

Correct	Incorrect
A	B
C	
D	
E	

We have had some students who simply refused to believe that it was possible to get credit for an incorrect choice; others will tell you that they are looking for the incorrect response but then revert to finding a "correct" answer because that response pattern has been so ingrained that they are simply unable to modify their conceptual framework. Either answering pattern will result in significantly lower scores.

Another all-too-common mistake that students make when encountering a I-B question is to read "except" as "accept." Our experience is that about 5 or 10 percent of students make that deadly mistake, which kills any chance to do well on a test. You can confirm this for yourself by simply asking students to decode the word "except" in a test question. Do not be surprised when you start seeing suggested meanings of "take" or "receive." It would be almost impossible to understand or answer a I-B question correctly, given this working definition. The confusion can be corrected in just a few minutes with couple of practice questions.

It is common for a standardized test to have about 20 percent of the questions fall into this I-B category. This is a great number of questions to miss because of a failure to understand one word.

Structural Type II: The Continuum

We think this type of question causes the most trouble because the answers do not fall into clearly correct and incorrect categories. Instead, they are placed along a scale. The basic diagram looks like this:

Getting credit for one of these questions requires knowing that a continuum is being used and then placing the choices along that scale. Here is a worked example:

Question: If $-7 \leq X \leq 7$ and $0 \leq Y \leq 12$, what is the greatest possible value of $Y - X$?

A. 0

B. -7

C. 7

D. 19

E. 5

Thus, by using a Type II scale to evaluate the appropriateness of each option, the correct response (D) is found.

We have students start by placing response A in the middle of the scale, no matter what it says. Then each of the following alternatives is placed on the scale by comparing it to the alternatives already placed *and* to the labels on the ends of the scale. We also encourage them to draw these diagrams on the test as they go, to deal with the questions more successfully.

Though common across many tests, the frequency of the Type II question is often determined by the subject being tested. Subjects such as math and science tend to have fewer questions of this type, while subjects such as reading have more. The choices do not fall into right and wrong, but rather lie on a continuum from best to worst. If a student approaches a question of this type looking for the single "correct" answer, confusion may result and getting credit is less likely. Often two or even three choices are quite close in meaning and are difficult to place on the Type II scale.

A student who knows and has practiced these three types of questions has an advantage because he or she has the ability to sort out answer choices through the application of the diagrams. One student wrote to us after passing her PRAXIS: "I listened to everything you told me, did my practices, studied differently and during the test used the graphs I, II and III" (personal communication, January 14, 2005). Many of our students have passed simply because they learned to deal more effectively with questions after studying the three types. They apply this powerful technique to the test questions during the test because we encourage them to diagram the answers as they go. Surely, this takes more time, but an increased percent correct is a big payoff. Increased reading speed can offset the increased time spent diagramming questions. That is, you can spend time learning to read faster so that those minutes gained can be invested in decoding the meaning of questions and their structures.

Good test takers construct models of test questions in their heads without conscious effort, but poor ones do not. Thus, teaching this technique to the latter will allow them to improve scores significantly.

Sometimes test makers seem to go out of their way to make questions more structurally complex than necessary. One ETS test sample question has remained with us for years. In this question the phrase "Least likely not to" was used. A Type II question is clearly indicated, but the answer desired is on the "Most likely" side rather than the "Least likely" side. Thus, "Least likely not to" and "Most likely" are equivalent. The diagram would look like this:

Most likely   Least likely

|⎯⎯|

*

Too many test takers misinterpret this question and think that it is asking for the "Least likely" alternative. They then go on to miss the question, not because they don't know the material, but because they have gotten lost in the complex verbiage of the question.

STRATEGY OF "ELIMINATION"

We often hear students tell us that they use or have been instructed to use the elimination technique. Too often this is not well applied or well thought out. When we follow up one of these statements with a few questions, the general weakness of that approach becomes evident. The assumption behind this technique is that questions have a single correct answer (see Type I-A above). While this is often true, it is not universal and its systematic application leads to disappointing test scores. For example, students tell us they are taught to eliminate wrong answers. What happens in the case of the "EXCEPT" question (Type I-B above) when credit is given for the wrong answer? Similarly, how does this strategy deal with a question where all of the answers are to some degree correct (Type II above)? A main idea question would be a good example of the latter. Here the choices are clearly not one correct response with four choices that are absolutely false, but an array of choices ranging from most correct to least correct. Going with the mind-set that one choice is absolutely correct will hinder any student from dealing efficiently and correctly with standardized tests.

Advocates of this elimination technique and students applying it are misleading themselves that they have a viable, useful technique. Any preparation program that advocates this technique is probably also deficient in other serious ways.

BLOOM'S TAXONOMY AND TEST PREPARATION

Another helpful way of looking into the content of most standardized tests is through the perspective of Bloom's Taxonomy. An understanding of the taxonomy and its ramifications can help students focus their study more productively.

The levels of Bloom's Taxonomy, going from least complex to most complex, are **K**nowledge, **C**omprehension, **A**pplication, **A**nalysis, **S**ynthesis, and **E**valuation. These levels are often thought of as a scale going from the least complex to the most complex, like this:

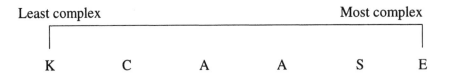

Because most tests consciously focus on assessing student understanding, objective questions are commonly set at the Comprehension or Application levels. Where students often go wrong is in expecting questions to be on the Knowledge level, since that had been sufficient to get them through classes and courses up to that point in their educational careers. We often have students come to us with high grade point averages who do terribly on standardized tests. We believe the possible problem is that when given material they can learn it and give back Knowledge-level answers with great facility, but without ever really understanding. Then when a question is posed that requires understanding, they fail because they have little experience in learning at that level.

An illustrative math example may help.

Example: On a math test you are presented with a complex expression like this:

$$\frac{(x^3y - y^3x)}{(x^2y - xy^2)}$$

It *appears* that the task required is one of factoring, and in fact four of the choices are algebraic simplifications. However, in small print off in the corner is the notation "denominator = 0." This changes everything. Now the question is not concerned with algebraic manipulation by factoring, but instead seeks to ascertain whether you realize the correct answer is "undefined," which is the last choice. This understanding also enables the test taker to move quickly to the next question with no calculation. In this case, realizing what the question is really asking enables the test taker to move on quickly after having answered the question correctly.

Mathematics is unique among the various subjects. It is unique because after the earliest grades, when the focus is on the "math facts," almost all math is taught at the Application level of Bloom's Taxonomy. It is exactly this characteristic that attracts some and repels others. Some truly poor math teachers manage to "push down" the subject to the Knowledge level and get students to do well, but these students will continue to do poorly on standardized tests for that same reason. If students work on fractions, then they have a test on adding fractions, work some more and have a test on subtracting fractions, and so on. They learn a pattern and can follow it faithfully to pass a test. If later they are given a test that has questions on percentages, word problems, geometry, the number line, and fractions, the students do poorly because they haven't been given practice on a wide range of math concepts together.

Another way test makers gauge a test taker's level of understanding is to present a table or graph and then ask a student to interpret it. The only way to prepare is to understand the principles, and students may have little or no experience doing that. Or, stated differently, understanding is the key, not simply knowing.

TEST CONTENT VOCABULARY: WHAT THE WORDS MEAN

As all tests have content, they also have vocabulary. Grammar tests expect students to recognize and correct specific grammar errors. For example, most grammar tests will expect that students deal with subject-verb agreement errors. To do that effectively and accurately, a student needs to comprehend the following vocabulary: subject, verb, number, singular, and plural. That same student may also need to understand person, tense, and conjugation. Without these concepts, the only approach is to decide if an answer "sounds" correct.

In the same vein, math tests have their own vocabulary. For example, area, perimeter, and pi, as well as symbols and formulae.

Thus, students need to learn to develop lists of words, terms, and symbols relevant to a particular test. We recall a success story from one of our preparation classes. We were asked to help a young lady prepare for the PRAXIS test in science. Her primary preparation focus was on vocabulary. She worked on it from two perspectives, general vocabulary words and specific science words. We supplied the former and she the latter. After passing the test, she wrote us that after her arduous study program, she was able to say that there were *no* words on the test that she did not know, none. You will understand the extent of her effort when you realize that she had to collect and learn an extensive science vocabulary that included biology (both zoology and botany), chemistry, physics, geology, astronomy, meteorology, and oceanography! Quite an accomplishment in eight weeks.

3

The Nature of Numbers

INTRODUCTION

The purpose of this chapter is to lay a foundation of numerical understandings that undergird all of the following chapters. The discussion will cover three points of view concerning numbers: mathematical, measurement, and finally Jean Piaget's thoughts on the growth of basic mathematical concepts.

RELEVANT CONCEPTS FOR ALL TESTS

Mathematicians look at numbers as a hierarchy of concepts. Each one is more inclusive than the previous one. They could also be conceived of as a series of concentric circles rather like a bull's-eye target.

Let's start at the center with the simplest and least inclusive concept.

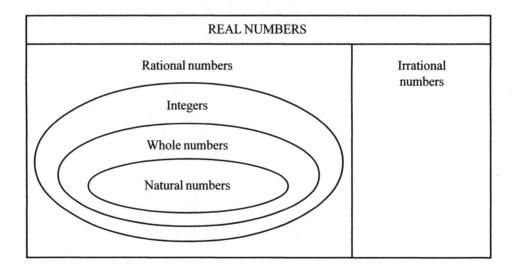

Natural numbers: These are the "counting" numbers that we start with as young children—1, 2, 3, and so on. Zero is not included. (Note: This term may occur in a constructed response question, where you would be asked to construct a graph or a number series. Think positive numbers and not zero.)

Whole numbers: These are the natural numbers above, but zero has been added. So this new concept goes: 0, 1, 2, 3, etc.

Integers: These are the whole numbers above, with the addition of negative numbers. This gives us –3, –2, –1, 0, 1, 2, 3, etc. Some subcategories of this concept are *positive integers*, which is simply a renaming of the whole numbers above, and *negative integers*, which are simply –1, –2, –3, and so on.

Rational numbers: These are numbers that can be expressed as a fraction (proper, improper, decimal, etc.) with a non-zero denominator. They include *integers*, which can be expressed as fractions. Rational numbers include all *terminating* decimals like $3/8$, which equals .375, and repeating decimals like $2/3$.

$$5/1 = 5 \quad 2/3 = 0.6666 \quad 3/8 = .375 \quad -16/2 = -8 \quad 5.8 = 5^8/_{10} = {}^{58}/_{10}$$

(Note: This means that integers are rational; they are whole numbers. Additionally, integers are used as components to make fractions.)

Irrational numbers: These are conceptually located in the "spaces" between rational numbers. These *cannot* be expressed as the ratio of two numbers (a fraction). In calculations, they are non-repeating, non-truncating (ending) decimals. Therefore, these numbers cannot represent a particular point on a number line. They have to be approximated.

$$\pi \text{ (pi)} \approx 3.1415..., \sqrt{2} \approx 1.414213562..., \sqrt{3} \approx 1.732050808...$$

All of the above represent *real numbers*, numbers that can be represented on a *number line*.

Imaginary numbers: These are numbers that are even roots of negative numbers. They take the form of $\sqrt{-1}$, which is represented as "i." For example, $\sqrt{-36} = \sqrt{36} \times \sqrt{-1} = 6\sqrt{-1} = 6i$. This abstruse concept allows many otherwise impossible calculations to take place.

Complex numbers: Are all possible sums of real and imaginary numbers in the form a + bi, where a and b are real and i ($\sqrt{-1}$) is imaginary. These could be plotted on a complex plane where the X axis is real and the Y axis is imaginary.

How Imaginary Numbers Work

Based on the power rule, when two negative integers are multiplied their product will become positive; therefore, their roots will be a positive radicand ($-8 \times -8 = 64$, or $\sqrt{64} = \pm 8$). A radical expression has two solutions, one negative and one positive. However, to keep both solutions real, the negative solution cannot be expressed under the radical $\sqrt{-64}$; the correct form is $-\sqrt{64}$ and $+\sqrt{64}$. Notice that the negative sign is outside of the radical. However, there are times when a negative *even power* is needed in order to make an equation hold true.

Example: Given the equation $x^2 + 1 = 0$

How can we make this a true statement? Would substituting -1 for x make it true?

$-1^2 + 1 = 0$ 	No, because the integer rule says negative • negative = positive: $-1^2 = 1$ and $1 + 1 = 2$, not 0.

What if we substitute zero for x, making the equation $0^2 + 1 = 0$? Would this make this statement true?

$0 + 1 = 1$ 	No, because zero squared is still zero and $0 + 1 = 1$, not 0.

Now, coming back to the same question, "How can we make this statement true?" there is a way to derive a solution but it will not involve real numbers—numbers that can be placed on a number line.

So a new type of number, i, can be substituted for x^2 to make the equation true:

$i^2 + 1 = 0$

Imaginary numbers evolved to explain negative numbers of *even* roots. The letter i, also written as $\sqrt{-1}$, is used as the imaginary *unit* (a precisely fixed quantity used to count or measure). It also follows that $i^2 = -1$. Since 1^2 is equivalent to -1, then $-1 + 1 = 0$, yielding a true statement.

So from this unit other powers of i can be defined:

$i^3 = i \cdot i^2$
$i^3 = i \cdot -1$ (remember, i^2 is equivalent to -1)
$i^3 = -i$ (i is i, and $i \cdot -1$ is $-i$)
$i^4 = i^2 \cdot i^2$ (is really $-1 \cdot -1$)
$i^4 = 1$ ($-1 \cdot -1 = +1$)

From here we can say that every 4 i's will equal 1.

Example: $i^{12} = (i^4)^3 = 1^3 = 1$ (How to think: Every 4 i's = 1)
Because i^4 is i • i • i • i   (this can be rewritten as i^2 times i^2)

$i^2 \cdot i^2$ (by definition i^2 equals –1)

–1 • –1   (–1 times –1, by the integer rule, equals 1)
1
$1^3 = 1$ To complete the example, 1^3 using the power rule is 1 • 1 • 1, which equals 1.

Example: $i^{39} = i^{36} \cdot i^3 = (i^4)^9 \cdot i^3 = 1^9 \cdot (-i) = -i$ (How to think: Every 4 i's = 1)

$$\begin{array}{r} 9 \\ 4\overline{)39} \\ 36 \\ \hline 3 \end{array}$$

How many 4s are there in 39? 9 with 3 remaining. So there are 9 groups of 4 i's and every 4 is equal to 1. There are 3 leftover i's. Therefore, two of the leftover i's are equivalent to –1. This leaves only one i left, which equals i. Now from here we simplify the power of $i^{39} = 1 \cdot -1 \cdot i = -i$. (*Note*: Any higher power of i's can be simplified by factoring all i's by 4 then by 2.)

$i^4 = 1$
$i^2 = -1$
$i = i$ or rewritten $\sqrt{-1}$

A second way to look at numbers is from the *measurement point of view*, that is, the underlying use made of numbers. The acronym NOIR will help you remember these.

Nominal: These are numbers that are only used to categorize or label and don't really function in a numerical sense. Example: categorizing male = 1 and female = 2. There is no sense that 2 is more than 1, or that 2 is twice as great as 1. The mode is the appropriate measure of central tendency to use here.

Ordinal: These numbers order values when nominal numbers do not, but the intervals between the numbers are not equal. That is, 2 is more than 1 and 4 is more than 3, but the "distance" between them is not always known to be equal. There is also no true zero. That is, zero does not represent an absence of the quantity. The order of horses finishing a race is ordinal. They would show 1st, 2nd, 3rd, etc. However, there is no assurance that there was an equal "distance" between 1st and 2nd or between 2nd and 3rd. A median would be the only appropriate measure of central tendency.

Interval: These have the characteristics of ordinal numbers but with the addition of equal intervals but no true zero. Temperature scales like Fahrenheit and Celcius are of this type. That is, 0° in either scale *does not* represent an absence of molecular movement but is just a point on the scale that extends further into the negative range. Means may be calculated here.

Ratio: Finally, we have the addition of a true zero. The Kelvin temperature scale is an example of this because 0° Kelvin (absolute zero) indicates the point where no molecular motion exists and thus no further reduction in temperature is possible.

Order of Operations

Strongly related to the nature of numbers is the nature of the way we must deal with numbers, which is embodied in the order of operations. Certain conventions have been adopted to assure that the same result will always be obtained in calculations. The following order is applied sequentially from 1 to 6, moving from left to right. Students often use the mnemonic device "Please Excuse My Dear Aunt Sally" to recall the order. This ordering is not absolute in that multiplication and division are performed together from left to right. Addition and subtraction are treated in the same manner, calculated as they occur from left to right.

1. Parentheses
2. Exponents (powers) and roots
3. Multiplication
4. Division
5. Addition
6. Subtraction

Example:

$$6 + 4 \cdot 3$$
$$6 + 12$$
$$18$$

Correct

Not

$$6 + 4 \cdot 3$$
$$10 \cdot 3$$
$$30$$

Incorrect

Multiplication is performed before *addition*

The order of operations is not performed correctly because *addition* is performed before *multiplication*

Example: $180 \div 6(12 + 8) \div 2 - 20 \div 5$

Answer:

$180 \div 6(12 + 8) \div 2 - 20 \div 5$
$180 \div 6(20) \div 2 - 20 \div 5$
$30(20) \div 2 - 20 \div 5$
$600 \div 2 - 20 \div 5$
$300 - 4$
296

1. Add what is in the parentheses and rewrite the expression.
2. Divide before multiplying because division comes first moving from left to right; then rewrite the expression.
3. Now multiply 30 times 20 because this time multiplying comes first moving from left to right.
4. Divide 600 by 2.
5. Divide 20 by 5 because division comes before subtraction; then rewrite the expression.
6. Finally, subtract 4 from 300.

JEAN PIAGET

Jean Piaget (1896–1980) was a widely respected Swiss investigator of the development of cognitive processes, with special focus on children up to the mid-teen years. As his writings have been translated into English and his students have become more influential in English-speaking countries, his theories have become widely accepted.

One of Piaget's main concerns was the acquisition of a sense of number by a child because without this deep understanding, all math becomes a frustrating application of poorly comprehended algorithms. His theory is usually classified as "constructionist" because he viewed persons, usually children, as active in the construction of numerical or logico-mathematical understanding. This is in opposition to the empiricist view that a child is a "blank slate" or "tabula rasa" upon which externals, people, and environment create conceptions of reality.

Piaget believed that number concepts *must* be created by a child and are not able to be "given" or "taught" by an adult or teacher. Let's consider the simplest situation. Two beads are identical in every respect. The concept of "two" or the "twoness" of the beads is created by the observer and is not in any way a property of the beads. This concept is the basis of addition (1 + 1 + 1 = 3) and by extension subtraction, multiplication, and division.

A corollary thought to the above on Piaget's concepts concerning the growth of "logico-conceptual" knowledge is what happens when this *does not* happen. Lacking this conceptual basis, it would seem likely that children, and then the adults they become, would not be able to function easily with more advanced mathematical concepts, regardless of the exceptional skills of the teacher or the superiority of the teaching materials. It is also possible that remediation would never be very successful until the fundamental conceptual acquisition takes place.

4

Number Line/Place Value

INTRODUCTION

A common type of problem on standardized tests simply asks about the relative position of numbers on the number line. The problem may be in any form: fractions, decimals, signed numbers, or complex expressions. There may be a calculation possible, such as conversion of fractions to decimals for ease of comparison, but all in all test makers simply wish to see whether you can locate numbers on a number line.

Any real number on a number line will always be greater than the number to its left and less than the number to its right because numbers get larger to the right and smaller to the left on a number line.

Example: Since "a" is on the left side of "b" it is less than "b" ($a < b$) and by the same token, since "b" is on the right of "a" it is greater than "a" ($b > a$).

RELEVANT CONCEPTS FOR ALL TESTS

The questions may take some of the following forms:

1. Is greater than ($>$)
2. Is less than ($<$)
3. Between (means it does not include the two end numbers)
4. Order from least to greatest (ascending order)
5. Order from greatest to least (descending order)

Following is an example of "is greater than":

Number Line/Place Value 23

Question: A pattern requires 1⁷/₈ yards of a certain fabric. If five remnants of fabric are as follows, which of these lengths will provide enough fabric and result in the least waste?

1¹/₂ yards   1³/₄ yards   1¹⁵/₁₆ yards   2¹/₂ yards   2¹/₁₆ yards

Answer: The correct answer will be the length just longer than 1⁷/₈. The first two are less than 1⁷/₈ and can be eliminated because they are too short. 1⁷/₈ is also equal to 1¹⁴/₁₆, so the piece 1¹⁵/₁₆ is just a bit (¹/₁₆ yard) longer. The last two are longer still and would have more waste. Thus the answer is 1¹⁵/₁₆ yards.

Picture this question using a number line.

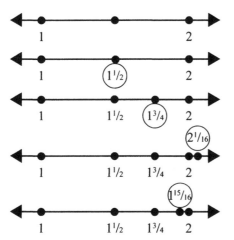

1. Label the number line 1 and 2 yards.

2. Now locate the yardage that will be halfway between 1 and 2 on the number line (1¹/₂ yards).

3. Now locate the fraction that will be halfway between 1¹/₂ and 2 (1³/₄ yards).

4. The only three choices left are 1¹⁵/₁₆, 2, and 2¹/₁₆.

We know that 1¹⁵/₁₆ is only ¹/₁₆ from 1⁷/₈, but 2 and 2¹/₁₆ are yet farther away, so 1¹⁵/₁₆ is the answer.

Remember, the above example is a section of a number line. The section includes whole numbers between 1 and 2.

Recall that there is an infinite number of smaller rational numbers (fractions or decimals) that exist between each whole number. The interval between 0 and 1 will only have fractional parts or the equivalent-form decimals less than 1. The interval between 1 and 2 will have the whole number 1 and the added fractional part or the equivalent decimal, depending on how you decide to divide the fraction (¹/₂, ²/₃, ¹/₄, ⁷/₈, ¹/₁₆, etc.). The interval between 2 and 3 will continue the same process except the whole number added to it will be 2 instead of 1.

Following is an example of "is less than":

Question: Which of the following is less than ⁵/₉?

⁵/₈   ²¹/₃₆   ²⁵/₄₅   ⁵⁵/₁₀₀   .565

Answer: Since fractions are often difficult to compare and one answer is already in decimal form, we can convert the rest to decimals by dividing the numerator by the denominator (this process is explained in Chapter 12). This results in:

$5/8 = .625$
$21/36 = .583$
$25/45 = .555$
$55/100 = .550$
$.565 = .565$

And since $5/9 = .555$, then

.625 is greater
.583 is greater
.555 is equal
.55 is less
.565 is greater

The answer is $55/100$.

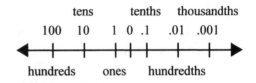

The first digit to the right of the decimal has the largest value. When comparing decimals, compare each place to determine the solution to the question being asked.

Here the question is asking for a value that is less than .555.

If it is difficult to see the values, follow the strategy below:

1. Circle the tenths place and compare which tenths place is the smallest. All are the same except for .625, so mark it off.

2. Now circle the hundredths place and compare which hundredths place is the smallest. It is between .555 and .55 because .565 is too large.

3. From here it is easy to determine that the solution is .55 because it can be thought of as .550 since a trailing zero after the decimal has no value but can be used to help compare its value.

Following is an example of "between":

Which of the following is between .50 and .60?

$4/5$ $2/5$ $7/14$ $6/11$ .601

Answer: If you convert the fractions to decimals you find:

$4/5 = .80$ too large
$2/5 = .40$ too small
$7/14 = .50$ not between

$^6/_{11} = .545$ between
$.601 = .601$ too large

Following is an example of "least to greatest":

Place the following in order from least to greatest:

$^3/_{10}$ $^3/_{14}$ $^3/_4$ $^3/_7$ $^3/_{12}$

Answer: If all numerators (top numbers) are the same, the largest denominator (bottom number) will be the smallest value. This gives:

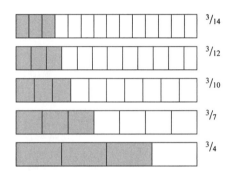

Notice that as the denominator gets smaller, the portions get larger. So the larger the denominator, the smaller the portions will be. Since the numerators are all the same, it is easy to see the order from least to greatest when the total areas are equal.

Or, all can be converted to decimals and the result is the same.

$^3/_4 = .75$ Greatest
$^3/_7 = .43$
$^3/_{10} = .30$
$^3/_{12} = .25$
$^3/_{14} = .22$ Least

Place Value

Often involved in this type of question is place value. That is, as we write numbers on both sides of the decimal, what are the names and values associated with those positions?

First, looking to the left of the decimal, we have the following:

IHG,FED,CBA.

A = Ones or units (10^0)
B = Tens, our basic unit in the base 10 number system (10^1)
C = Hundreds (10^2)
D = Thousands (10^3)
E = Ten Thousands (10^4)
F = Hundred thousands (10^5)

G = Millions (10^6)
H = Ten millions (10^7)
I = Hundred millions (10^8)

Then, looking to the right of the decimal, we have:

.ABCDE

A = tenths (A/10) Note: 10 = 10^1
B = Hundredths (AB/100) Note: 100 = 10^2
C = Thousandths (ABC/1,000) Note: 1,000 = 10^3
D = Ten thousandths (ABCD/10,000) Note: 10,000 = 10^4
E = Hundred thousandths (ABCDE/100,000) Note: 100,000 = 10^5

Even though we use the base 10 number system, other systems exist. Let's look at one system common to computer systems, the binary or base 2 system. It is composed of only two numbers, 0 and 1, which makes it useful in electronic circuits or switches because they can be either on or off (1 or 0). Here is the format of the binary system:

IHG,FED,CBA.

A = Ones (2^0 = 1)
B = 2 (the base) (2^1 = 2)
C = 2^2 (4)
D = 2^3 (8)
E = 2^4 (16)
F = 2^5 (32)
G = 2^6 (64)
H = 2^7 (128)
I = 2^8 (256)

Note that these have exact correspondence to the same place value in the base 10 number system above.

Now let's count with binary numbers.

Base 10	Binary	Base 10	Binary
1	1	6	110
2	10	7	111
3	11	8	1000
4	100	9	1001
5	101	10	1010

Example: The binary number equal to 21 in the base 10 is:

A. 10101
B. 11000
C. 11100
D. 10001
E. 11111

A. $16 + 4 + 1 = 21$
B. $16 + 8 = 24$
C. $16 + 8 + 4 = 28$
D. $16 + 1 = 17$
E. $16 + 8 + 4 + 2 + 1 = 31$

1. Write down the base you want to convert to as the power of that base. Here the question is to change from base 10 to base 2.
   So I write down the powers of 2:
   $2^0 = 1 \quad 2^1 = 2 \quad 2^2 = 4 \quad 2^3 = 8 \quad 2^4 = 16$
   $ 1 \quad 0 \quad 1 \quad 0 \quad 1$
   This is all that we need, since $2^4 = 16$, which is the greatest number that will divide into 21.

2. Now descend through each power asking the question:
   Will 16 divide into 21? Yes, one time, with a remainder of 5. Put 1 under 2^4.
   Will 8 (which is 2^3) divide into 5? No, so put a 0 under 2^3.
   Will 4 (which is 2^2) divide into 5? Yes, one time, with a remainder of 1. So put 1 under 2^2.
   Will 2 (which is 2^1) divide into 1? No, so put a 0 under the 2^1.
   Will 1 (which is 2^0) divide into 1? Yes, one time, so put a 1 under 2^0.

This gives 10101 as the answer.
Check: A. $16 + 4 + 1 = 21$ Correct

Example: The binary number 1010 equals what number in the base 10 system?

A. 102
B. 10
C. 2
D. 16
E. 202

1. Note how many places the binary number extends (four places).

2. Write base 2 four places out and place each number underneath each power in reverse.
   $2^0 = 1 \quad 2^1 = ② \quad 2^2 = 4 \quad 2^3 = ⑧$
   $ 0 \quad 1 \quad 0 \quad 1$

3. Add the value of each power that has a one underneath it.
   $2^1 = 2 \quad 2^3 = 8$

These are the only powers that have ones; all of the other powers have zero.
   Correct answer: B, because $1010 = 8 + 2 = 10$.

Stem and Leaf Plots

Stem and leaf plots are related to the content of this chapter as well as the chapter on statistics (Chapter 14). Below is a stem and leaf plot that presents data in a graphic manner. The units digit is separated and ordered from the rest of the places (tens, hundreds, etc.)

Example: Display the following grades in a stem and leaf plot: 95, 98, 71, 74, 87, 90, 80, 81, 79, 82, and 85.

Answer:

tens Stem	ones Leaf
7	1 4 9
8	0 1 2 5 7
9	0 5 8

Notice that in the stem column the tens place of each number is represented only one time and in the leaf column each of the units ascends numerically and is categorized by tens (the stem).

Thus, the stem of 7 denotes 7 tens and the 1, 4, and 9 denote units. This represents the following three numbers: 71, 74, and 79.

It is also possible to calculate means, medians, and modes from data presented in this manner. In the above problem, for example, the median is 82 because five numbers are lower and five numbers are higher.

A stem and leaf plot can show so much in the way of analyzing data. Here we can see how many grades are in the nineties, eighties, and seventies. It indicates whether instruction was a success or if reteaching is needed, and it indicates whether grades are evenly distributed. It can also tell whether a gross mistake was made in grading or if a student had a bad day—grades that are way out of range are called outliers. See Chapter 21 for more information on analyzing graphs.

END OF PRAXIS I, 0014, 0511, AND 0146

In the more complex tests, the tasks are identical to the preceding but the relationships are more complex.

Example: Which of the following statements is true?

$3/8 < 4/11 < 5/13$
$4/11 < 3/8 < 5/13$
$5/13 < 4/11 < 3/8$
$4/11 < 5/13 < 3/8$
$3/8 < 5/13 < 4/11$

Answer: Since all the choices involve only three fractions to evaluate, convert each to a decimal.

$^3/_8 = .375$
$^4/_{11} = .364$
$^5/_{13} = .385$

Thus, the correct order is $^4/_{11} < ^3/_8 < ^5/_{13}$, which is the second choice.

Example: If a = .99, which of the following is/are less than a?

(Remember that the variable "a" is the one being squared, not the variable "X." The variable X is the coefficient to the "a.") Also, when we do not know what the coefficient is, it is understood to be 1, and 1 times anything is itself. Usually the 1 is not shown, just as a decimal at the end of a whole number is not shown but is understood (21 = 21). When it is needed we can make it visible for calculations or clarity.

$$\underbrace{\sqrt{a} \times X\sqrt{a}} = a \text{ (Where X = 1)}$$

The X is 1 and 1 times the square root of "a" equals $\sqrt{a}$. Any time you square a square root, you simply drop the radical. In essence, squares undo square roots.

I. $\sqrt{a}$
II. a^2
III. $1/a$

A = none   B = I only   C = II only   D = III only   E = II and III

Answer: First assign numerical values to the alternatives.

I = .995 (.99 = .995 × .995)
II = .98 (.98 = .99 × .99)
III = 1.01 (1.01 = 1/.99)

This yields only II as less than .99, which makes C the best answer.

Note: When we take the square root of the decimal .99, the solution is larger than the original value and when we square the decimal .99, the value is smaller. Why? Multiplying whole numbers yields larger products; our minds are used to seeing it that way. However, when multiplying by a number less than 1, the opposite occurs. Here .99 is less than 1. When you multiply two fractions or their equivalent decimals, the product will be smaller, and squaring a number is simply multiplying. Now, why does the decimal increase when taking the square root of a decimal? Because it is the inverse of squaring a number, which means that it undoes squaring.

Now, what about dividing any whole number by a number less than 1? Division by a number less than 1 will produce a larger quotient because you will have more groups with smaller portions. When you divide by a whole number or by a whole number attached to a decimal, you will have fewer groups but larger portions. Remember that the divisor tells the dividend how many groups to divide the dividend into.

Now you should be able to use mental math to obtain a reasonable solution.

1. Let "a" represent the same value in all situations. Since it is less than 1, squaring that decimal will yield a number smaller than the square root. So, choice I is discarded.
2. Dividing a whole number by a decimal will always yield an answer greater than 1. Choice III can now be discarded.
3. For any number less than 1, its square will be even smaller. Choice II has to be the solution.

Example: If $0 < x < 1$, which of the following lists the numbers in increasing order?

A. $\sqrt{x}, x, x^2$
B. $x^2, x, \sqrt{x}$
C. $x^2, \sqrt{x}, x$
D. $x, x^2, \sqrt{x}$
E. $x, \sqrt{x}, x^2$

Realize that the relationship described by $0 < x < 1$ is saying that the question deals only with numbers between 0 and 1. Understanding this fact is an absolutely vital step in getting the correct answer because numbers in this range react differently than numbers larger than 1 or negative numbers. It also rules out 0 and 1, which also have different characteristics.

Answer: Substitute a number, say .5. Then $x^2 = .25$, $\sqrt{x} = .707$, and of course $x = .5$. This gives B as the correct answer (.25, .50, and .707). Be careful in answering questions like this because at first glance it seems counterintuitive for the square of a number to be smaller than the number itself.

Understanding how decimals behave will allow you to use only mental math to solve the problem above. On a test this means saving valuable time for more complicated examples.

1. Think about what the example $0 < x < 1$ is saying: x is greater than 0 but less than 1. So what types of numbers could x represent? Any decimal or fraction between 0 and 1 will satisfy the conditions.
2. What is the example asking you to do? Order the decimal values from least to greatest.
3. Now use what you know about the behavior of decimals when squaring or

taking the square root. Squaring a decimal yields a smaller decimal and calculating the square root undoes squaring, which gives a larger solution.

4. Discard any solutions that do not have x^2 as a possible first number. A, D, and E are out.

5. Now think about it. The initial decimal will have to lie between the square and the square root. So B is correct.

Example: Which of the following CANNOT be expressed as the sum of three consecutive integers?

18   24   28   33   36

Answer: Three consecutive integers can be represented by x, x + 1, and x + 2. Notice that the example does not tell us the initial value. All it does tell us is that the values are consecutive, which means one after another. Since the initial value is unknown, it may be represented with a variable. Here "x" is used. Then I know that the next integer will be one more than the initial value (x + 1) and the next will be one more than the previous value (x + 2). From here you simply add the three expressions to get 3x + 3. To find the initial integer you must undo the addition by subtraction and the multiplication by division. This means to take each of the options above and subtract, then divide. Notice that the process is the reverse order of operations when going backwards to find the initial integer. But here the question wants the *wrong* answer. So make sure to use Type I-B answering strategy when seeking the wrong answer. See Chapter 2 for an explanation.

Shortcut: If the sum of the three numbers must conform to the pattern 3x + 3, then the sum itself must be divisible by 3. Only one answer is not divisible by 3, and that is 28. It is the answer.

Longer approach: The sum would be 3x + 3. For a number to be the sum of three consecutive integers it must therefore be divisible by 3 after 3 has been subtracted. Now let's look at the contenders:

18 – 3 = 15 and 15/3 = 5   Could be the sum.
24 – 3 = 21 and 21/3 = 7   Could be the sum.
28 – 3 = 25 and 25/3 = 8$^1/_3$ Could NOT be the sum.
33 – 3 = 30 and 30/3 = 10   Could be the sum.
36 – 3 = 33 and 33/3 = 11   Could be the sum.

Analysis: The correct answer is 28.

Beware: Since four of the answers are even and one (33) is odd, don't choose 33 because it is different. The issue is not odd/even numbers but whether you understand divisibility.

5

Decimals

INTRODUCTION

Decimals are around us all of the time. They are most evident in money. For example, nearly every advertisement features a price—2 for $1.49, 3 for $10.00, and so on. You may also see decimals in ads concerning interest rates. For example, home loans at 4.75 percent, 5.25 percent, and so on. However, there is a world of decimals beyond the two places commonly used in currency and interest rates.

RELEVANT CONCEPTS FOR ALL TESTS

Addition and Subtraction of Decimals

The operations of addition and subtraction with decimals are closely related here, as they are in the chapter on fractions (see Chapter 10). However, here the rule is a bit different. Instead of finding a common denominator, you must *align the decimals*. It is this process that allows you to add or subtract "like" terms; that is, tenths are added to tenths, and so on. Not aligning the decimals would mean that you were attempting to add or subtract "unlike" terms.

Example: 49.003 + 29.7 =

Answer: You do *not* set up as:
```
   49.003
+    29.7
   49.300  INCORRECT
```

But rather this way:
```
   49.003
+ 29.700
   78.703  CORRECT
```

Decimals

(Zeroes may be added to the right as needed without changing the value of the number.)

However, if you think about the simplest case with addition and subtraction of whole numbers, you will realize that we *actually do* align the decimals. They are always in the same place and usually not even written.

Example: 146 + 29 is really 146.0 + 29.0 (or 146. + 29.)

Answer: So 146   and 146.   and 146.0
             + 29   + 29.   + 29.0
              175   175.   175.0

All are equivalent.

Multiplication of Decimals

Here there is no need to align the decimals—you merely have to sum the number of digits to the right of the decimal in the two factors being multiplied. Then you place the decimal in the product that many places from the right. Why? Because of place value.

Example: 21.45 × 10.4

Answer: 21.45 ⎤ 3 digits to right
        × 10.4 ⎦ of decimals
          8.580   tenths place multiplied by each above
         00.000   ones place multiplied by each above
        214.500   tens place multiplied by each above
        223.080 }   3 digits to right of decimal

See that each row moves one place to the left using the Power of Ten. You have been doing this on simpler calculations without actually knowing and applying the rule.

Example: What is the cost of 5 boxes of detergent at $4.99 each?

Answer: $4.99 ⎤ 2 digits to right
        ×   5 ⎦ of decimal
         $24.95 } 2 digits to right of decimal

Division of Decimals

Long division with decimals gives many test takers their greatest problem. The rule is to move the decimal on the divisor (the number you are dividing *by*) as far to the right as possible and then correspondingly move the decimal on the dividend (the number being divided) the same number of places to the right, adding zeroes if necessary.

Example: $46.4 \div 3.2 = ?$

Answer: $3.2\overline{)46.4}$

Becomes $32.\overline{)464.}$ when both decimals are moved one place to the right. This yields the answer of 14.5. You work out the actual steps to get 14.5. Remember, $.32\overline{)4.64} = 3.2\overline{)46.4} = 32.\overline{)464.} = 14.5$.

Place value is the underlying issue here!
We can demonstrate this by using a technique called "partial quotient."

$3.2\overline{)46.4}$ | 12.5 ← 1. How many 3.2's are in 40?
           1.875 ← 2. How many 3.2's are in 6?
         + .125 ← 3. How many 3.2's are in .4?
           14.5

$.32\overline{)4.64}$ or $3.2\overline{)46.4}$ or $32\overline{)464}$

have identical quotients because of place value.

SUMMARY OF PROCEDURES

Addition and subtraction: align the decimals, then proceed to add or subtract.

Multiplication: sum the number of digits to the right of the decimal in the two factors being multiplied. Then place the decimal in the product the same number of digits obtained above from right to left.

Division: Move the decimal all the way to the right in the divisor and correspondingly move the decimal in the dividend the same number of places to the right (place value). Place the decimal in the quotient directly above the decimal in the dividend. Be careful about this placement because incorrect answer choices will involve faulty decimal placement in the answer. For example, answer choices may include .47, 4.7, and 47.

6

Rounding

INTRODUCTION

Questions involving rounding are common on many tests, and all require a prior knowledge of place value. In general, there are two ways rounding can be included in questions:

1. Rounding a number to a given place value. Example: Round to the nearest hundredth or round to the nearest tenth.
2. Rounding to do either mental math or estimating.

Here is an important thought on rounding/estimating. These two complementary processes should come into play in answering *any* math question! As test takers are presented with virtually any question, the first reaction should *not* be to start calculating or pressing calculator keys. Too many students fail to perform the vital task of estimating an answer before beginning to calculate. Because of this omission virtually any answer will be accepted, even if it makes no sense at all. An experienced math teacher went so far as to say that students had no business calculating until after they had estimated an answer. Previous use of slide rules promoted this estimating process because with that venerable instrument, an estimate was absolutely necessary since the user always had the obligation to place the decimal! Persons initially trained with a slide rule made the estimation process second nature, even after calculators became common. This retained trait kept many of those students from accepting erroneous answers.

RELEVANT CONCEPTS FOR ALL TESTS

To perform the first function, a test taker must know place value. See Chapter 4 for a review.

Process

You will be told the place to be rounded.

Step 1. Look to the digit that is immediately to the right of the digit indicated. If it is 5 or greater, add 1 to the specified digit and substitute zeroes for the deleted digits.

Step 2. If it is 4 or less, substitute zeroes for all digits to the right.

Example: Round 1,278,634 to the nearest thousand.

Answer: Let's consider the 8, which is in the "thousands" place. To decide about rounding, look at the first digit to the right of the 8, which is 6. Since 6 is greater than 5, the 8 will increase to 9. Zeroes are placed to the right of the 9 to keep the number in its numeric value.

1,278,634 → 1,279,000

Example: Round 1,248.634 to the nearest hundredth.

Answer: Let's consider the 3, which is in the "hundredths" place. The first digit to the right is 4, which is less than 5, so the answer is:

1,248.634 → 1,248.630 or 1,248.63

(Notice that it does not change the value to drop the trailing zeroes.)

It is also possible to use rounding in a second way, which is in estimating an answer.

Example: The dimensions of a room are 11 feet wide, 13 feet deep, and 9 feet high. What is the approximate number of cubic feet in that room?

Answer: Using approximations, we would get 10 × 13 × 10. This makes a simple calculation of 100 × 13 = 1,300 cubic feet.

The calculated value is actually 1,287 cubic feet, which is within 13 cubic feet of the estimation. However, if the question asks for an estimate, the approximation (here 1,300) is correct and the exact answer (1,287) is incorrect.

Here are a few more to do:

Given: 1,235,741.86492

Example: Round to the nearest:

A. Hundred thousand
B. Thousand
C. Hundred
D. Ten
E. One
F. Tenth
G. Hundredth
H. Thousandth

Answers:
A. 1,200,000
B. 1,236,000
C. 1,235,700
D. 1,235,740
E. 1,235,742
F. 1,235,741.9
G. 1,235,741.86
H. 1,235,741.865

7

Signs

INTRODUCTION

Careful attention to signs is often the key to passing a mathematics test. Avoidable errors are often only the result of carelessness and lack of attention.

RELEVANT CONCEPTS FOR ALL TESTS

Signed numbers (+ and –) indicate direction on the number line. These signs are not to be confused with the identical signs that are used to indicate the operations of addition and subtraction.

To explain how integers and signs work, let's start with a number line.

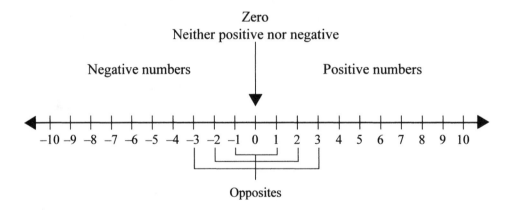

To the right of zero are the natural numbers. Each natural number has an opposite. The set of natural numbers and their opposites plus zero creates a new set of numbers, called integers. Look at the number line above and see how each number to the right of zero corresponds to a number to the left of zero. This continues infinitely in both directions.

From the number line you can see that any *negative* number is less than any *positive* number, and that zero is less than any *positive* number but greater than any *negative* number.

In order to understand how to compute integers, it is important that you understand the concept of *absolute value*. The notation is a number between two vertical bars: for example, |5|. We know that absolute value can never be negative. How is this related to a number line? When we add or multiply integers on a number line we are really moving distances, and distance is never negative. For example, when a car goes in reverse, it does not go a negative distance. Although the car is going backwards, it is going a positive distance. This is the same concept on a number line.

Adding with a Number Line

Rule: To add numbers having the same sign, add the absolute value of the numbers. Give the result the same sign as the numbers being added.

The following is an example of adding two positive integers:

$$3 + 5 = 8$$

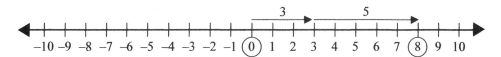

Add the positive numbers 3 and 5 by first starting at zero and following the arrow three units to the right. Then, from the right end of this same arrow, follow the second arrow five units to the right. The answer is below the second arrow.

Note: The absolute bars are not necessary. Both integers are positive because we are moving eight positive units to the right.

The following is an example of adding two negative integers:

$$-4 + -2 = -6$$

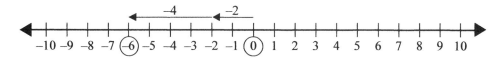

Add −2 and −4 on a number line by first starting at zero and following the arrow two units to the left. Then, from the left end of the first arrow, follow the second arrow four additional units to the left. The answer is below the second arrow.

Another way to see two positive integers added:	Another way to see two negative integers added:
If I have 3 positive chips and I add 5 positive chips to them, I will have 8 positive chips. *The sum of two positive numbers is positive.*	If I have 2 negative chips and I add 4 negative chips to them, I will have 6 negative chips. *The sum of two negative numbers is negative.*
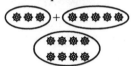	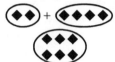
Making 8 positive chips	Making 6 negative chips

Adding Numbers with Different Signs

Rule: To add numbers with different signs, find the absolute value of the numbers and subtract the smaller absolute value from the larger. Give the answer the same sign as the number with the larger absolute value.

The following is an example of adding integers with different signs:

$$-1 + 6 = +5 = 5$$

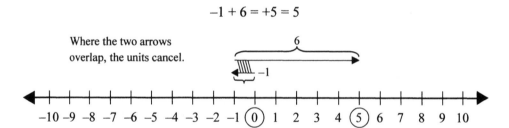

The arrow starts at zero and moves one unit to the left. From the left end of the first arrow, another arrow is drawn six units to the right. This is a little tricky because we are adding opposites, (−1) to +6, which cancels one of the units. Here we find the absolute value of both integers and then subtract the smaller absolute value from the larger absolute value (6 subtract 1 = 5). The sign is determined by which number is the largest. In this case 6 is larger than 1. The sign of 6 is positive, so the sum will be positive.

The following is another example of adding integers with different signs:

$$-5 + 3 = -2$$

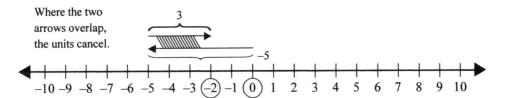

Where the two arrows overlap, the units cancel.

Now let's add another negative integer and a positive integer. Notice that both arrows are completely on the negative side of the number line. The arrow starts at zero and moves five units to the left, and then the second arrow starts at the end of the first arrow and moves three units to the right. This is a little tricky because we are adding five negative units (–5) to three positive units, which cancels three units (–5 + 3 = –2).

The same procedure is used as in the first example, finding the absolute value of each and then subtracting the smaller value from the larger value. The sign of the larger value will determine the sign of the answer.

Another way to see different signs added is making pairs with a zero sum.

Make 3 pairs with a zero sum.

If I have 1 negative chip and I add 6 positive chips to it, I will have 5 positive chips.
$$-1 + 6 = +5$$

If I have 5 negative chips and I add 3 positive chips to it, I will have 2 negative chips.
$$-5 + 3 = -2$$

These two chips cancel each other because 1 + –1 = 0, leaving a total of 5 positive chips.

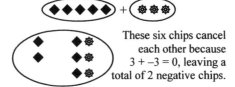

These six chips cancel each other because 3 + –3 = 0, leaving a total of 2 negative chips.

Subtracting with the Number Line

Rule: First, change the subtraction symbol to addition and change the sign of the subtrahend. Then subtract the smaller absolute value from the larger. Give the answer the same sign as the number with the larger absolute value.

Minuend – subtrahend = difference

The following is an example of subtracting signed numbers:

$$8 - 12 = -4$$

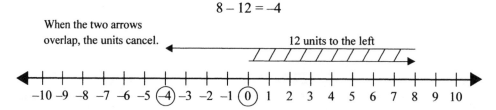

The arrow starts at zero and moves eight units to the right. From the right end of the first arrow, another arrow is drawn 12 units to the left. This is a bit tricky because we are adding eight units to –12 units, which in reality cancels eight units. See below.

$$\text{Example: } 8 + (-12) = -4$$

Note: Using the *addition rule*, the absolute bars are helpful because one of the integers is negative. So here we find the absolute value of both integers and then subtract the smaller absolute value from the larger absolute value (12 minus 8 = 4).

The sign is determined by which number is the largest. In this case 12 is larger than 8. The sign of 12 is negative, so the sum will be negative, in this case –4.

Note: Using the *subtraction rule*, first change the operational sign to addition and then change the sign of the subtrahend to its opposite: $8 + (-12) = -4$. Then follow the addition rule.

The following is another example of subtracting signed numbers:

$$-8 - (-2) = -6$$

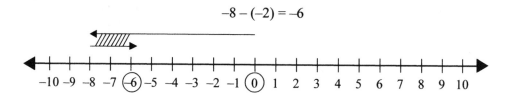

The arrow starts at zero and moves eight units to the left. From the left end of the first arrow, another arrow is drawn two units to the right. This is a little tricky because we are subtracting negative 2 units from –8 units. This equates to adding 2 units, negating two units.

$$\text{Algebraic example: } -8 + 2 = -6$$

Here we find the absolute value of both integers and then subtract the smaller absolute value from the larger absolute value (8 subtract 2 = 6).

The sign is determined by which number is the largest. In this case 8 is larger than 2. The sign of 8 is negative, so the sum will be negative, in this case –6.

Note: Using the *subtraction rule*, first change the operational sign to addition and then change the sign of the subtrahend to its opposite: $-8 - (-2) = -6$. Then follow the addition rule.

Signs

Note: It is important not to confuse an operation sign for a positive or negative sign. The four basic operations are $+$, $-$, $\div$, and $\times$. When an integer does not have a sign, it is understood to be positive.

$-8 + 9 = 1$	$-8 + (-9) = -17$	$-8 - 9 = -17$	$8 - (-9) = 17$
The operation is addition. The 9 is a positive integer.	The operation is addition. The 9 is a negative integer.	The operation is subtraction. The 9 is a positive integer. Here, follow the subtraction rule by changing the operational sign and the subtrahend to its opposite.	The operation is subtraction. The 9 is a negative integer. Here, follow the subtraction rule by changing the operational sign and the subtrahend to its opposite.

Word problems will not always have the operation stated specifically, so it is important that you properly translate the meaning of words or phrases that will indicate subtraction.

Words or Phrases	Numerical Expressions from Word Phrases	Examples of How to Translate Word Phrases
Subtracted *from*	$20 - 14 = 6$	Means 14 subtracted from 20
Decreased by	$7 - (-5) = 7 + 5 = 12$	Means 7 decreased by -5
Less *than*	$2 - 3 = 2 + (-3) = -1$	Means 3 less than 2
Difference between	$-5 - (-9) = -5 + 9 = 4$	Means -5 plus -9
Minus	$-10 - 6 = -10 + (-6) = -16$	Means -10 minus 6

Other words that may indicate subtraction include "below," "underneath," "descend," "drop," "borrowed," "owe," "down," and "lose."

IMPORTANT: READ EACH WORD PROBLEM CAREFULLY.

Notice above that any time *from* or *than* is used as part of a phrase, the numeric expression is transposed. So as part of a numeric expression in a word problem, you should do the same.

Following are additional translated mathematical phrases:

5 subtracted from the sum of 9 and -2

Here the word "sum" indicates that addition is also part of the expression. Notice that "sum" precedes the numbers 9 and -2, so this means to add these two numbers. Since 5 precedes the expression "subtracted from," the 5 should be subtracted after 9 and -2 are added. Remember the clue word "from." This is also a hint that 5 is to be written last: $[9 + (-2)] - 5 = 2$. From this point the order of operations should be performed.

6 less than -2

Remember the clue word "than," which indicates to reverse the order; however, if you read carefully you will know that 6 cannot be less than -2, thus the expression should be translated as $-2 - 6$. Now write the numeric expression using the subtraction rule: $-2 + -6 = -8$, or $-2 + (-6) = -8$. See that "6 less than -2" is different from "6 *is* less than -2." The expression "6 *is* less than -2" indicates $6 < -2$, which is a false statement. The important word "is" changes the meaning of the whole numeric expression.

Multiplying Positive Numbers Using a Number Line

Rule: Multiplying two positive factors will have a positive product. The following is an example of multiplying signed numbers:

$$3 \cdot 2 = 6$$

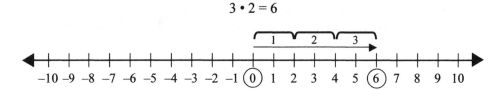

Starting at zero, the arrow moves three groups of two units to the right and ends at 6, the product of 3 times 2.

Note: The line above indicates a positive distance of 6 in three groups of two units on the positive side of the number line. This will result in a positive product.

The following is an example of multiplying using different signs on a number line.

Rule: When multiplying a positive number by a negative number or vice versa, the product will always be negative.

$$3 \cdot -2 = -6$$

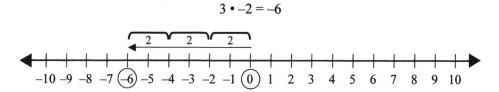

Starting at zero, the arrow is moving three groups of –2 units to the left and ending at –6, which is the product of 3 times –2.

Note: We are moving a distance of three groups to the negative side of the number line. This will result in a negative product.

Word problems will not always say that the operation is multiplication. So it is important that you can properly translate the meaning of words or phrases that will indicate multiplication.

Words or Phrases	Numerical Expressions from Word Phrases	Examples of How to Translate Word Phrases
Product *of*	$3 \cdot -5 = -15$ or $(3)(-5) = -15$	The *product of* 3 and –5
Percent *of*	$.06 \cdot 20 = 1.20$ or $(.06)(20) = 1.20$	6 percent *of* 20
Of (used with fractions)	$1/3 \cdot 6 = 2$ or $(1/3)(6) = 2$	$1/3$ *of* 6
Times	$-5 \cdot -7 = 35$ or $(-5)(-7) = 35$	–5 times –7
Twice (means two times)	$2 \cdot 4 = 8$ or $(2)(4) = 8$	Twice 4

Following are additional translated mathematical phrases:

| The product of 3 and the difference of 6 and –2 | 20 percent of the sum of 7 and –5 |

Note: "Of" occurs twice. Only one refers to multiplication. Read carefully for content. Here the 3 is multiplied by the difference of 6 and –2.

$$3[6 - (-2)] = 24$$
$$3(6 + 2) = 24$$

(Remember the subtraction rule.)

Here 20 percent is changed to a decimal (.20) and then multiplied by the sum of 7 and –5.

$$.20[7 + (-5)]$$
$$.20(2) = .40$$

Follow the order of operations. (Clear parentheses, then multiply.)

Division of Positive and Negative Numbers on a Number Line

Rule: Dividing a positive number into a positive number will result in a positive quotient.

The following is an example of dividing using positive numbers:

$$6 \div 2 = 3$$

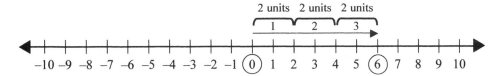

Starting at zero, the arrow is moving three groups of two units to the right and ending at 6. This time the number line reflects the dividend (6), the number of units represents the divisor (2), and the total number of brackets represents the quotient (3).

The following is an example of dividing using different signs on a number line.

Rule: Dividing a negative number by a positive number will result in a negative quotient.

$$-8 \div 4 = -2$$

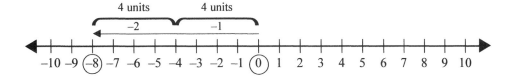

Starting at zero, the arrow is moving two groups of four units to the left and ending at –8. This time the number line reflects the dividend (–8), the number of units represents the divisor (4), and the total number of brackets represents the quotient (–2).

Multiplying and Dividing When Both Numbers Are Negative

To model multiplying and dividing integers, it may be better to use chips than a number line.

$$-6 \div -2 = +3$$

There are six chips that happen to be negative.

There are two groups of six negative chips that have three units in each group.

When multiplying or dividing where both integers are negative numbers, think about how many chips are there after multiplying or how many chips are in each group when dividing.

Here is a graphic that can be used to find the correct sign when multiplying or dividing *only*:

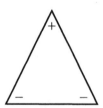

Look at the two adjacent vertices for the signs you are using when multiplying or dividing with signs. Then look at the opposite side to find the resulting sign for the solution.

Example: –6 times or divided by –7

Step 1: Look at the two vertices that have those signs. The two bottom vertices are negative, negative.

Step 2: Find the vertex opposite those two vertices and you will see that your answer will have a positive sign.

Addition The following relationships hold true:
+ (+6) = +6 = 6
+ (–6) = –6
+4 + (+3) = +7 = 7
+4 + (–3) = + 1 = 1

Subtraction	These are also true: $-(+6) = -6$ $-(-6) = +6 = 6$ $+4 - (+3) = +1 = 1$ $+4 - (-3) = +7 = 7$
Multiplication	These are a bit more complicated: $(+6)(+6) = +36 = 36$ $(-6)(+6) = -36$ $(+6)(-6) = -36$ $(-6)(-6) = +36 = 36$
Division	Like multiplication, these are also somewhat complicated: $+6 \div +6 = +1 = 1$ $+6 \div -6 = -1$ $-6 \div +6 = -1$ $-6 \div -6 = +1 = 1$
Powers	Signs are relevant in raising to powers: $+3^2 = +9 = 9$ $-3^2 = +9 = 9$ $+3^3 = +27 = 27$ $-3^3 = -27$ (Remember, $-3 \cdot -3 \cdot -3 = -27$)
Roots	Signs are even more relevant in determining roots: $\sqrt{16} = \pm 4$ [because $(+4)(+4) = 16$ and $(-4)(-4) = 16$] $\sqrt[3]{-27} = -3$ [because $(-3)(-3)(-3) = -27$] $\sqrt{-16} = \sqrt{16 \cdot -1} = 4\sqrt{-1} = 4i$ where $I = \sqrt{-1}$

This last entry brings us finally into the realm of imaginary numbers, which are actually beyond the scope of tests 0014, 0511, and PRAXIS I Math.

Example: In California, the highest point is Mt. Whitney, at 14,494 feet above sea level. The lowest is Death Valley, at 282 feet below sea level. What is the difference between the two?

A graphic will help solve this example:

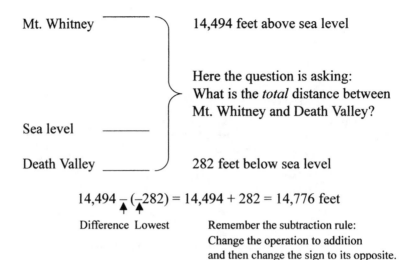

14,494 feet above sea level

Here the question is asking:
What is the *total* distance between Mt. Whitney and Death Valley?

282 feet below sea level

$$14{,}494 - (-282) = 14{,}494 + 282 = 14{,}776 \text{ feet}$$

Difference   Lowest

Remember the subtraction rule:
Change the operation to addition
and then change the sign to its opposite.

The total distance between the two locations is 14,776 feet, not 14,212 feet, which will be given as one of the answer choices!

What is:

Example: $-5 \times (-6) = ?$

Answer: $-5 \times (-6) = -5 \times -6 = +30 = 30$

Example: $-3 + (-3) = ?$

Answer: $-3 + (-3) = -3 - 3 = -6$

Example: $+5 + (-10) = ?$

Answer: $+5 + (-10) = +5 - 10 = -5$

END OF PRAXIS I, 0014, 0511, AND 0146

Example: $2|x + 3| - 9 = 5$. Solve for x.

Answer:
$$\begin{aligned}
2|x + 3| - 9 &= 5 \\
+9 &+ 9 \\
\hline
2|x + 3| &= 9 + 5 \\
\frac{2|x + 3|}{2} &= \frac{14}{2} \\
x + 3 &= 7 \\
-3 &- 3 \\
\hline
x &= 4
\end{aligned}$$

8

Fundamental Operations: Addition and Subtraction

INTRODUCTION

A number of questions will include no more than addition and/or subtraction. It is assumed that test takers have memorized all the facts and can perform these operations quickly and accurately without the use of a calculator.

Many test takers overrely on calculators and start pushing buttons too soon. They also fail to *think* about a solution and don't have an approximate answer in mind before starting to calculate. This can be a serious mistake because *any* answer will then be accepted.

RELEVANT CONCEPTS FOR ALL TESTS

Fundamental Properties of Addition and Subtraction

Commutative Law: The order that numbers are added doesn't change the sum. $(2 + 3 = 3 + 2)$

Identity for Addition: The addition of zero does not change the value. $(8 + 0 = 8)$

Identity for Subtraction: The subtraction of zero does not change the value. $(8 - 0 = 8)$

Associative Law for Addition: The grouping of numbers to be added does not change the result. $(8 + 2) + 7 = 8 + (2 + 7)$

See that subtraction is neither commutative nor associative, because rearranging the numbers will not result in the same answers.

Example: $5 - 3 = 2$ whereas $3 - 5 = -2$ (but 2 does *not* equal -2)

Subtraction and addition are *inverse operations*. This means that they undo each other. See the following table at column 2, row 6 and column 5, row 7.

$2 + 4 = 6 \qquad 5 + 2 = 7$
$\downarrow \qquad\qquad\qquad \downarrow$

	-	0	1	2	3	4	5	6	7	8	9
	0	0									
	1	1	0								
	2	2	1	0							
	3	3	2	1	0						
	4	4	3	2	1	0					
	5	5	4	3	2	1	0				
$6 - 4 = 2 \rightarrow$	6	6	5	4	3	2	1	0			
$7 - 2 = 5 \rightarrow$	7	7	6	5	4	3	2	1	0		
	8	8	7	6	5	4	3	2	1	0	
	9	9	8	7	6	5	4	3	2	1	0

ANOTHER IMPORTANT CONCEPT: PLACE VALUE

Ones are added or subtracted from ones, tens are added or subtracted from tens, hundreds are added or subtracted from hundreds, and so on.

Two Types of Subtraction Problems

$$132 - 22 = 100 + 30 + 2$$
$$- (20 + 2)$$
$$100 + 10 + 0 = 110$$

$$132 - 41 = \overset{130}{\cancel{100} + \cancel{30}} + 2$$
$$- (40 + 1)$$
$$90 + 1 = 91$$

Here just simply subtract. No regrouping (borrowing) is necessary.

1 hundred + 3 tens + 2 ones
<u> − 2 tens − 2 ones</u>
1 hundred + 1 ten + 0 ones = 110

Regrouping is necessary here. You regroup by breaking down the *one* hundred into bundles of tens. This means 10 groups of ten. Since it is *only* one hundred, the whole hundred will be regrouped into 10 bundles of ten in the tens place before subtracting 4 tens from 3 tens. The 3 tens are added to the 10

Fundamental Operations: Addition and Subtraction   51

tens, making 13 tens. Now take 4 tens away.

(If 2 had been in the hundreds place, only one of the hundreds would have been regrouped into bundles of 10 and the other would have remained in the hundreds place.)

$$\begin{array}{r} 0 \text{ hundreds} + 13 \text{ tens} + 2 \text{ ones} \\ - 4 \text{ tens} - 1 \text{ ones} \\ \hline 0 \text{ hundreds} + 9 \text{ tens} + 1 \text{ ones} = 91 \end{array}$$

Two Types of Addition Problems

$$\begin{array}{r} 241 + 22 = 200 + 40 + 1 \\ + 20 + 2 \\ \hline 200 + 60 + 3 = 263 \end{array}$$

Here just simply add. No regrouping (carrying) is necessary.

$$\begin{array}{r} 2 \text{ hundreds} + 4 \text{ tens} + 1 \text{ ones} \\ + 2 \text{ tens} + 2 \text{ ones} \\ \hline 2 \text{ hundreds} + 6 \text{ tens} + 3 \text{ ones} = 263 \end{array}$$

$$242 + 58 = \overset{3}{\cancel{200}} + \overset{5}{\cancel{40}} + 2$$
$$+ 50 + 8$$
$$\overline{300 + 0 + 0 = 300}$$

Regrouping is necessary here. Since our numeration system is based on 10, the ones place must be regrouped because the sum of the ones place is 10, which equals one bundle of 10 (2 + 8 = 10). We bundle the ten from the ones place and move it to the tens place because of place value. Now, we have zero ones and 10 tens and 2 hundreds.

$$\begin{array}{r} 2 \text{ hundreds} + \cancel{4} \text{ tens} + 2 \text{ ones} \\ + 5 \text{ tens} + 8 \text{ ones} \\ \hline \downarrow \end{array}$$

$$\begin{array}{r} 2 \text{ hundreds} + 5 \text{ tens} + 0 \text{ ones} \\ + 5 \text{ tens} + 0 \text{ ones} \\ \hline \end{array}$$

Now do the same process in the tens place. Notice that now we have 5 tens instead of 4 tens in the top expression. When we add the tens places, we have 10 tens, which means that we really have 1 hundred because 10 • 10 = 100. Now we have to regroup again by first changing the 10 tens to its equivalent of 1 hundred and placing it in the hundreds place, and placing zeroes in the tens place.

What to Conclude from This

The test makers are interested to know whether you understand subtraction or addition, not just the process of deriving a solution.

- They want to know that you *understand* that when you subtract and add, you are working with place value, which is really powers of ten. This may be a constructed response question that really wants you to elaborate place value by showing or telling what you did and why you did it.
- They want to know if you *understand* that subtraction and addition are inverses of each other. They may set up a situation that will test

(continued)

> your understanding of how to solve a problem using the inverse strategy.
> - They want to know if you *understand* why subtraction is not commutative or associative, and why addition is. Your job is to elaborate the explanation so they know you understand.

3 hundreds + 0 tens + 0 ones
———————————— 0 tens + 0 ones

The sum after regrouping is:

3 hundreds + 0 tens + 0 ones = 300

Some word problems, though fairly complex in presentation, are relatively simple to solve.

Example: A comet appeared in 1905. In what years will it make its next three visits if it reappears every 75 years?

Answer: 1905 + 75 = 1980
         1980 + 75 = 2055
         2055 + 75 = 2130

Thus, the years for the comet's reappearance are 1980, 2055, and 2130.

Example: Salesmen for a newspaper are expected to sell the following number of ads each week:

Between 12 and 18 quarter-page ads

Between 9 and 12 half-page ads

Between 7 and 12 full-page ads

What are the most and least number of total ads that salesmen are expected to sell?

Answer:

Least = 12 + 9 + 7 = 28
Most = 18 + 12 + 12 = 42

They are expected to sell between 28 and 42 ads per week.

9

Fundamental Operations: Multiplication and Division

INTRODUCTION

A number of questions will include no more than multiplication and/or division. It is assumed that test takers have memorized all the facts and can perform these operations quickly and accurately without the use of a calculator.

Many test takers overrely on calculators and start pushing buttons too soon. They also fail to *think* about a solution and have an approximate answer in mind before starting to calculate. This can be a serious mistake because any answer will then be accepted. A talented math teacher once said that students have no business starting any calculation if they have no approximate idea of what the answer should be. He felt that estimation should precede calculation.

RELEVANT CONCEPTS FOR ALL TESTS

Fundamental Properties of Multiplication

Commutative Law: Changing the order of factors does not change the product (e.g., 2 • 4 = 4 • 2). Notice that neither the shape nor the size of the arrays changes. The only thing that changes is the position. Both have an area of 8 and a perimeter of 12.

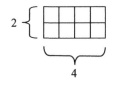 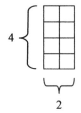

Identity for Multiplication: Multiplying by 1 does not change the value. That is, you have only one group of 8 with a total of 8 elements ($8 \cdot 1 = 8$) (in contrast to the other two examples).

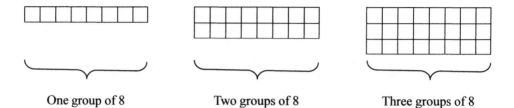

One group of 8   Two groups of 8   Three groups of 8

Multiplicative Property of Zero: Zero times any factor is zero. That is, to have zero groups of any factor is to have none ($2 \cdot 0 = 0$).

Associative Law for Multiplication: The grouping of numbers to be multiplied does not change the product. That is, $(8 \cdot 2) \cdot 6 = 8 \cdot (2 \cdot 6)$.

$$16 \cdot 6 \qquad\qquad 8 \cdot 12$$
$$96 \qquad\qquad 96$$

Distributive Law of Multiplication: This is more easily shown than strictly defined.

$5(2 + 3 + 4) = (5 \cdot 2) + (5 \cdot 3) + (5 \cdot 4)$ The Distributive Property
$\quad 5(9) \quad = \quad 10 + 15 + 20$ is used to multiply each term
$\quad\; 45 \quad = \quad\quad\; 45$ inside the parentheses.

OR

$8 \cdot 43 = 8 \cdot (40 + 3)$ Here it is used to spread out
$\quad = (8 \cdot 40) + (8 \cdot 3)$ or to separate the terms into
$\quad = \quad 320 + 24$ more manageable parts to
$\quad = \quad\; 344$ solve using mental math.

Here is a good example using the Commutative, Associative, and Distributive properties.

Here are four terms to simplify. They are as follows:

$2a^2 + b + a + 2ab$ Rearrange the terms using the Commutative Property.

Fundamental Operations: Multiplication and Division

$2a^2 + 2ab + a + b$	Now use the Associative Property to manipulate the expression even more.
$(2a^2 + 2ab) + (a + b)$	Now that the Associative Property is used, factoring can be done to simplify the expression.
$2a(a + b) + (a + b)$	Here the first binomial has a "2" and the "a" variable in common. Factor them outside the parentheses (factoring reverses the distributive property). In the second binomial there are no common factors.
$(2a + 1)(a + b)$	Use the Commutative Property again to rearrange the terms. Then use the Associative Property to group the factors outside the binomial. Write the binomials that are the same only once. This expression can now be solved easily with a little more information.

Example: If sound travels 1,100 feet/sec., how many feet will it travel in:

A. 30 seconds	Here seconds is the smallest unit, so simply multiply total feet times total seconds $1,100 \cdot 30 = 33,000$ feet
B. 2 minutes	Here the Associative Property can be used $(60 \cdot 2) \cdot 1,100 = 132,000$ feet
C. 1 hour	Here the Associative Property can be used $(60 \cdot 60) \cdot 1,100 = 3,960,000$ feet

Note: Units of time are as follows: 60 seconds is equivalent to one minute; $60 \cdot 60 = 3,600$ seconds, which is equivalent to one hour.

Fundamental Properties of Division

The components of a division expression can be expressed in three ways. Here the divisor tells the dividend how many groups there will be and the quotient tells how many will be in each group.

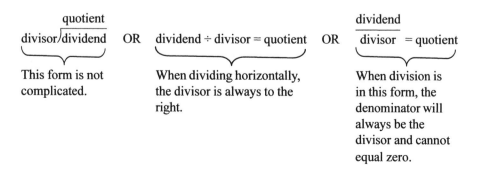

| This form is not complicated. | When dividing horizontally, the divisor is always to the right. | When division is in this form, the denominator will always be the divisor and cannot equal zero. |

Since multiplication and division are inverses of each other, then division by zero cannot be allowed. For example:

> If five times two equals ten (5 • 2 = 10),
> then ten divided by two equals five (10 ÷ 2 = 5).
> *This is a TRUE mathematical statement.*
> If ten times zero equals zero (10 • 0 = 0),
> then zero divided by zero *should* equal 10 if this
> mathematical statement is to remain TRUE.
> However, it does NOT equal ten (0 ÷ 0 ≠ 10).
> *This, then, is a FALSE mathematical statement.*

This example is a segue to the Division by Zero Rule.

Division by Zero Rule: Zero cannot be the denominator or the divisor. If it is, the answer will always be "undefined." Sometimes a zero is not readily seen. Beware of some binomial expressions in which the denominator equals zero. For example:

$$\frac{5}{4a - 4a}$$

4a – 4a will equal zero, which will be undefined, so no solution can be calculated.

The test makers assume that the test takers can perform both long and short division quickly and without error. Review and practice to make sure.

> Example: A car averages 15 miles per gallon of gasoline. How many gallons are needed to drive a distance of:
>
> A. 480 miles
>
> B. 525 miles
>
> C. 1,155 miles
>
> Answers:
>
> A. 480/15 = 32 gallons
>
> B. 525/15 = 35 gallons
>
> C. 1,155/15 = 77 gallons
>
> Example: At a bond sale, $225,000 worth of bonds was sold. How many bonds were sold if each bond was sold for:

A. $25

B. $50

C. $150

Answers:

A. $225,000/$25 = 9,000 bonds

B. $225,000/$50 = 4,500 bonds

C. $225,000/$150 = 1,500 bonds

One important type of division that is problematic is dividing when one or more zeroes are in the dividend. Students are not always sure how many zeroes should be part of the quotient. For example:

$$
\begin{array}{r}
104 \\
52\overline{)5{,}408} \\
\underline{52} \\
208 \\
\underline{208} \\
0
\end{array}
$$

- After dividing 52 into 54 there is a remainder of 2.
- 52 will not go into 2, so bring down a 0. This is where the problem begins.
- When there is a remainder and the divisor will not divide into it, then another digit must be brought down. If the divisor still won't divide into it, then this is when a zero is placed in the quotient and not before.
- After the zero is placed in the quotient, bring down the last digit. Although 52 did not divide into 2 or 20, it will divide into 208 four times, leaving no remainder.

Often in other mathematical tasks, such as the simplification of fractions, knowing the divisibility rules is imperative. These help you select the number that can be divided into another *evenly*. On any timed test this tool will be very helpful.

Divisible by	If
2	It is even (ends in 2, 4, 6, 8, or 0)
3	The sum of the digits is divisible by 3
4	The last two digits (read as a two-digit number) are divisible by 4
5	It ends in 0 or 5
6	It is even and the sum of the digits is divisible by 3
8	The last three digits (read as a three-digit number) are divisible by 8
9	The sum of the digits is divisible by 9
10	Ends in zero

Prime Numbers are numbers that have only two factors (1 and the number itself). Although prime numbers are seen as odd numbers, they are not by definition because two is an even number but is also prime. Thus prime numbers are not the product of smaller numbers. Two is prime because by definition it has only two factors, 1 and itself; therefore, two and three are the only two prime numbers that will be adjacent. Also, if you look at a series of prime numbers you will see that they generally become farther apart as the numbers increase in size. Between 1 and 99 there are 25 prime numbers; between 200 and 299 there are 16 and between 500 and 599 there are only 14. As numbers become larger, more small numbers are available as factors.

Numbers that are *not* prime are *composite* except for the number one. Composite numbers are numbers that have at least three factors. For example, 12 is composite because it has six factors, 1, 2, 3, 4, 6, and 12. An example of a composite number with only three factors is 4 (1, 2, and 4). When a factor is duplicated it is listed only once.

The number one is neither prime nor composite because it has only one factor, and that is 1. Remember that when a factor is duplicated it is listed *only once*. The concepts of prime and composite are important in solving problems because you can look at a product and know the factors, or that it is prime and no further modification is possible.

Knowing prime and composite numbers and the divisibility rules are important math strategies for ruling out test answers. Here are some worked examples:

Is the Example Divisible by	Example	
2	584	Yes, because the last digit is even (divisible by 2).
3	1,836	Yes, because the sum of its digits is a multiple of 3 (sum of the digits: $1 + 8 + 3 + 6 = 18$; $18/3 = 6$).
4	1,256	Yes, because the last two digits are divisible by 4 ($56/4 = 14$). (This is true because every hundred is divisible by four, so the last two digits are the issue.)
5	1,030	Yes, because the last digit ends in a zero ($1,030/5 = 206$).
6	1,836	Yes, because this number meets both conditions—the sum of the digits is a multiple of 3 and the last digit is even, so it will divide by 2 evenly. (First condition: $1 + 8 + 3 + 6 = 18$; $18/3 = 6$. Second condition: 1,836 is even.)
8	1,280	Yes, because dividing 8 into 280 results in no remainder ($280/8 = 35$; then 1,280 will divide by 8 evenly: $1,280/8 = 160$). (This is true because every thousand is divisible by eight, so the last three digits are the issue.)
9	173,457	Yes, because the sum of the digits is a multiple of nine (same as the divisibility rule for 3). The sum of these digits is 27 ($1 + 7 + 3 + 4 + 5 + 7$). Twenty-seven is a multiple of 9 ($27/9 = 3$ and no remainder): $173,457/9 = 19,273$.

**Is the Example
Divisible by   Example**

10   125,490   Yes, because the last digit is zero and so is divisible by 10.

Examples: The following are divisible by what number or numbers? (Note that some may be prime)

Number	Divisible by
100	2, 4, 5, and 10
2,001	3
1,998	9, 2, 3, and 6
6,402	2, 3, and 6
284,004	6, 2, and 3
847,296	8, 4, and 2
17,065	5
581	7
109	prime
547,214	2
70,587	9 and 3
6,203	prime
91	7 and 13

After perfecting an understanding of prime and composite numbers, you can find commonality between two factors, if any. These are strategies that are helpful in finding all the prime factors of a composite number and vice versa.

This example will take a composite number and factor it to its prime factors (prime factorization).

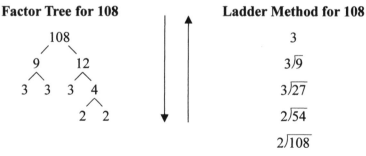

Factor Tree for 108   **Ladder Method for 108**

The factor tree tiers downward with composite factors until it is simplified to its prime factors, in ascending order (2, 2, 3, 3, 3).

The ladder method starts from the bottom and ladders upward, dividing only common prime factors into the composite number until the quotient is prime—notice the prime factors are the same (2, 2, 3, 3, 3).

The *Greatest Common Factor* (GCF) is the largest factor common in two numbers. Mastering this concept will help in the solution of ratios, simplifying fractions, and factoring. The two methods above are strategies used in finding the prime factorization of a number, and they also can be used to find the greatest common factor of two numbers.

Example: What is the GCF for 143 and 99?

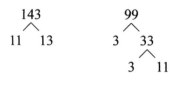

- Now list both sets of prime factors and compare.

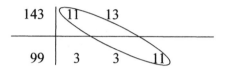

- The only common factor is 11; and since it is the only common factor, it is the greatest common factor.

Example: What is the GCF for 24 and 48?

- Write all common factors in the form of the example on the right.
- Circle all common factors.
- Write down each pair, then multiply them to get the greatest common factor.
- Notice that there is one factor left without a match. Discard it as part of the GCF.

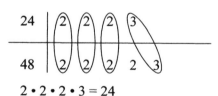

$2 \cdot 2 \cdot 2 \cdot 3 = 24$

GCF is a tool used in problem solving to associate things evenly in the smallest number of groups. For example:

Given 18 roses and 48 carnations to group in an arrangement that would utilize all of them without having any left over, use the GCF of 6 (2 • 3) and divide them into the two numbers to find how many are in each group. In the example on the previous page, there are three roses for every eight carnations, because 18/6 = 3 and 48/6 = 8.

Least Common Multiple (LCM). Also called *Least Common Denominator* (LCD). In order to understand LCM, it may be easier to model the subtle difference between GCF and LCM first.

Factors are the numbers that are multiplied to get a product. The product that you obtain can also be thought of as a multiple, depending on usage. For example:

factors product	factors multiples
5 • 6 = 30	1 • 3 = 3
	2 • 3 = 6
	3 • 3 = 9
	4 • 3 = 12
	etc.

GCF means finding the factors that both numbers have in common. It could be one factor or many factors, as we have seen in the above examples under GCF.	LCM means finding the smallest multiple that two or more factors will divide into evenly.
GCF is used most commonly in reducing fractions, finding equivalent fractions, or in factoring.	LCM is used most commonly in finding a common denominator when adding and subtracting fractions.

There are several strategies for finding the LCM. One is called the listing method. All the same strategies that are used in finding the GCF are also used in finding the LCM, with only slight differences. For example, the listing method consists of only the multiples of two numbers and then finding the first multiple that the two numbers have in common.

Sometimes this can be a long list and may not be advantageous to use; however, this is a great visual method.

Find the LCM of 7 and 6: 7 | 7 14 21 28 35 42 49 56 63 ... 126
                          6 | 6 12 18 24 30 36 42 48 54 ... 126

After writing several multiples of both 7 and 6, the lowest multiple that 7 and 6 have in common is 42. This means that 42 is smallest number or multiple that 6 and 7 will divide into evenly. The next multiples that 7 and 6 have in common are 84 and 126; 84 is 42 • 2 and 126 is 42 • 3.

The following is another example using the Factor Tree Method:

What is the LCM (LCD) of 100 and 88? (Hint: you are looking for the lowest multiple that may be divided by both numbers evenly.)

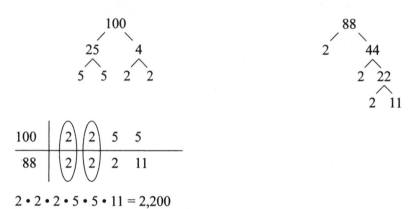

2 • 2 • 2 • 5 • 5 • 11 = 2,200

Notice the difference in finding the LCM when using the Factor Tree Method. When finding the GCF, the factors that are not paired are discarded. With the LCM, all pairs are made, and then what is left is written as well. This method prevents duplicating any pair of factors. Once the factors are written, multiply them to get the LCM. So 2,200 is the least common multiple that both 100 and 88 will divide into evenly. As said before, there are many more multiples that 100 and 88 will divide into, but they will not be the *least*, or smallest, multiple.

Challenge Example: What are the GCF and LCM for 12, 36, and 54?

Answer: 12 = 3 • 2 • 2
36 = 3 • 3 • 2 • 2
54 = 2 • 3 • 3 • 3

GCF = 2 • 3 = 6
LCM = 3 • 3 • 3 • 2 • 2 = 108

Following are word problems associated with LCM and GCF:

LCM Problem: A tourist train company has two rides leaving from the same station. The lake ride takes 15 minutes and the hill ride takes 18 minutes. If both trains leave the station at the same time, when is the next time they will be back at the station at the same time?

Fundamental Operations: Multiplication and Division

```
15 | 15 30 45 60 75 90      In 90 minutes, or 1 hour and 30 minutes
18 | 18 36 54 72 90
```

GCF Problem: The members of the boys' choir and girls' choir are marching in a parade. The teacher wants equal rows with the same number of boys and girls in each. There are 16 boys and 20 girls. What is one arrangement that satisfies the teacher's requirement?

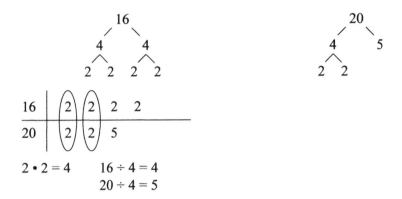

$2 \cdot 2 = 4$ $16 \div 4 = 4$
             $20 \div 4 = 5$

4 boys and 5 girls in four rows or 8 boys and 10 girls in two rows.

The former solution answers the GCF question with the largest number of groups; the latter solution satisfies the teacher's requirement with fewer groups.

10

Fractions: Addition and Subtraction

INTRODUCTION

Fractions represent a serious challenge for many students. In fact, it is not an exaggeration to say that weak math students nearly always exhibit a severe weakness in fractions. It is easy to spot. When given a problem, these weak students *avoid* the use of fractions. This avoidance costs points on a test because the use of fractions is often the most efficient route to an answer.

RELEVANT CONCEPTS FOR ALL TESTS

Before discussing how to add or subtract fractions, let's discuss what a fraction is. A proper fraction is a real number written as a ratio that represents a part of a whole or a set. Since a fraction is a real number it can be placed on a number line. Between each whole number there are fractional parts. For example, there is an infinite number of fractions between the whole numbers 5 and 6, which is only one unit on the number line:

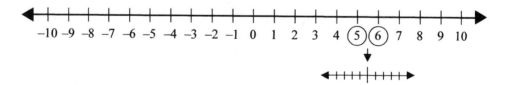

This model represents one unit, no matter what two numbers it is between, because a proper fraction is greater than zero but less than one whole. If I wanted to divide the unit between 5 and 6 into six equal parts, it could look like this:

Fractions: Addition and Subtraction 65

5 ¹/₆ ²/₆ ³/₆ ⁴/₆ ⁵/₆ ⁶/₆ } 6 over 6 equals 1. In this case, 5 + 1 = 6.

Divided into eight equal parts, my unit between 5 and 6 would look like this:

5 ¹/₈ ²/₈ ³/₈ ⁴/₈ ⁵/₈ ⁶/₈ ⁷/₈ ⁸/₈ } 8 over 8 equals 1. In this case, 5 + 1 = 6.

The point is, a unit can be divided into as many parts as desired between any two whole numbers.

Both addition and subtraction require the use of common denominators. Fractions must be put in a common metric where the denominators (bottom numbers) are the same before calculations may be made. For example:

⁵/₈ + ①/₄ =
⁵/₈ + ②/₈ = ⁷/₈

Here the denominators are different. So in order to add these two fractions, an equivalent fraction that carries the same value must be used. ¹/₄ and ²/₈ are equal but the parts are cut into smaller pieces so that the two numbers can be added. See the model below:

 First model: One fourth is shaded.

 Second model: I divide this model in half (eighths). I now have two eighths shaded. The size of the array did not change, only the size of the portions. This is what you do when finding a LCD.

 ⁵/₈ + ²/₈ = ⁷/₈. Notice that seven eighths are shaded.

Least Common Denominator: You need to find the smallest number that is evenly divisible by both denominators. As a last resort you can always use the product of the denominators, but it will make the calculation a bit cumbersome. Repeating the above example, using the product of the denominators, the computation would look like this:

⁵/₈ + ¹/₄ = ²⁰/₃₂ + ⁸/₃₂ = ¹⁴/₁₆ = ⁷/₈

8 • 4 = 32   Reduced to the lowest terms

²⁸/₃₂ is correct, but the best answer choice will be ⁷/₈ because that has been simplified.

Improper Fractions: If your calculation gives you an improper fraction (numerator larger than the denominator, e.g., $12/9$), you will need to pull out the whole numbers and then make sure the remaining fraction is also simplified. This can be done by dividing the denominator into the numerator (12 ÷ 9). 9 will go into 12 one time with three ninths left over. Three ninths can be simplified to one third. See below:

$$12/9 = 1\,3/9 = 1\,1/3$$

Reminder: *Fractions and division are inseparable!* $5/8$ and 5 divided by 8 are exactly the same.

Borrowing in Subtraction: When subtracting mixed numbers (like $1\,3/8$), it is often necessary to borrow one from the whole number and convert it to its fractional equivalent before proceeding to the actual subtraction. For example:

$2\,5/8 - 3/4 =$	First, find the common denominator for 8 and 4.
$2\,5/8 - 6/8 =$	6 cannot be subtracted from 5, so borrow a whole (1) from the two.
$1\,13/8 - 6/8 =$	Next, convert the 1 into a fraction that is equivalent to 1. Because any number over itself is equal to 1 and my denominator is 8, I can use $8/8$. $8/8 + 5/8 = 13/8$.
$1\,7/8$	Finally, add the two numerators together. Now subtract $6/8$ from $1\,13/8$.

Remember, $1 = 8/8$.

Word Problems with Fractions

Example: Find the total number of pounds in two bags if their weights are:

A. Each $14\,1/4$ lbs.

B. $15\,3/4$ lbs. and $20\,3/8$ lbs.

C. $125\,7/8$ lbs. and $250\,15/16$ lbs.

Answers:

A. $14\,1/4 + 14\,1/4 = 28\,2/4 = 28\,1/2$ lbs.

B. $15\,3/4 + 20\,3/8 = 15\,6/8 + 20\,3/8 = 35\,9/8 = 36\,1/8$ lbs.

C. $125\,7/8 + 250\,15/16 = 125\,14/16 + 250\,15/16 = 375\,29/16 = 376\,13/16$ lbs.

Example: Find the number of pounds left in a bag weighing 100 pounds if the number of pounds removed from it is:

A. $90^{1}/_{4}$

B. $75^{3}/_{8}$

C. $35^{87}/_{100}$

Answers:

A. $100 - 90^{1}/_{4} = 99^{4}/_{4} - 90^{1}/_{4} = 9^{3}/_{4}$

B. $100 - 75^{3}/_{8} = 99^{8}/_{8} - 75^{3}/_{8} = 24^{5}/_{8}$

C. $100 - 35^{87}/_{100} = 99^{100}/_{100} - 35^{87}/_{100} = 64^{13}/_{100}$
   Or, alternatively, $100 - 35.87 = 64.13$

Example: Normal body temperature is $98^{3}/_{5}$ degrees. How many degrees above normal is a temperature of:

A. 100 degrees

B. $101^{1}/_{2}$ degrees

C. $103^{3}/_{10}$ degrees

Answers:

A. $100 - 98^{3}/_{5} = 99^{5}/_{5} - 98^{3}/_{5} = 1^{2}/_{5}$ degrees

B. $101^{1}/_{2} - 98^{3}/_{5} = 101^{5}/_{10} - 98^{6}/_{10} =$
   $(100^{10}/_{10} + {}^{5}/_{10}) - 98^{6}/_{10} = 100^{15}/_{10} - 98^{6}/_{10} = 2^{9}/_{10}$ degrees

C. $103^{3}/_{10} - 98^{6}/_{10} = 102 \,({}^{10}/_{10} + {}^{3}/_{10}) - 98^{6}/_{10}$
   $102^{13}/_{10} - 98^{6}/_{10} = 4^{7}/_{10}$ degrees

END OF PRAXIS I, 0014, 0511, AND 0146

Often these other tests will format a fraction question combined with algebra:

Example: Simplify: $(x^2 + 2x)/x$

Answer: $(x^2 + 2x)/x$
  $x(x + 2)/x$
  $x + 2$

11

Fractions: Multiplication and Division

INTRODUCTION

Multiplication and division of fractions are fundamentally different from addition or subtraction (dealt with in the previous chapter) because you do not need to find a common denominator. In fact, really all that you do for both operations is multiply because to perform division, you simply invert (use the reciprocal) the second number (the divisor) and multiply.

RELEVANT CONCEPTS FOR ALL TESTS

Multiplying Fractions

The process is simple because you multiply the numerators (the top numbers) and then multiply the denominators (the bottom numbers). Then the only thing left to do is to simplify wherever possible.

Example: What is $1/3 \times 4/5$?

$4/5$

1. Here I have an array with four of five squares shaded to represent $4/5$.

$12/15$

2. Now divide the entire array into thirds to represent $12/15$, equivalent to $4/5$ because the portions are cut into smaller pieces but the area is the same.

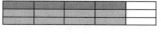
$1/3$ of $4/5$

3. Now slash one-third of the shaded array to represent $1/3$ of $4/5$. Another way to see this is: the slashed portions are the product (4 out of 15 smaller portions).

Answer: $\frac{1}{3} \times \frac{4}{5} = \frac{4}{15}$ or 4 out of 15

Notice that when we multiply fractions (part to part) or any *two* factors that are *less than 1*, the product of the denominators will be larger than the two original denominators but it really represents smaller portions, so the resulting fraction will be smaller because a larger denominator indicates a smaller value if the numerators are the same.

An example word problem would be:

The café has half of a pie left to sell, and the waitress sold $\frac{1}{3}$ of it to a customer. How much of the original pie did she slice for the customer?

The top half of the pie is the → part of interest.

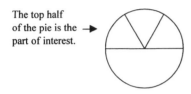

Just looking at this diagram without seeing the one below, you could easily assume that the waitress sliced $\frac{1}{3}$ of the pie, but if you read closely, the question is asking how much of the *entire* pie she sliced after half was sold.

In order to know the portion of the slices, we need to know how the *entire* pie is divided, which is in sixths.

In this case, the question is really asking for a portion of a portion. The equation would be $\frac{1}{3}$ of $\frac{1}{2} = \frac{1}{6}$. Remember, "of" means times.

So each slice is $\frac{1}{6}$, so she sold $\frac{1}{6}$ of the original pie.

Notice that the half is a bigger portion than $\frac{1}{6}$ *and* the $\frac{1}{3}$ is a bigger portion than $\frac{1}{6}$, yet the denominator of 6 is a larger digit than the denominators 2 and 3. *This is an important concept for test takers.* The test maker wants to know if you truly understand how fractions behave. They want to know whether you realize that the product in the *denominator* is how the parts of the whole were originally divided and the product in the *numerator* is the portion that you are seeking as a solution to the problem. They also want to know if you understand that when two proper fractions are multiplied, the product will be *smaller* than either of the original fractions, even though the digits in both numerator and denominator may be larger. This is because the area may not change but the size of the portions does. The smaller the portions, the more you have of them. The larger the portions, the fewer you have of them.

Sometimes whole numbers and fractions are multiplied. In this type of question we have a fraction multiplied by a whole, not part to part but part to whole.

Example: What is $\frac{1}{3}$ of 25? (remember, "of" indicates multiplication)

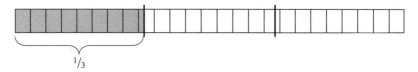

1. Here I have an array of 25.
2. I divide the 25 into thirds. Notice that in order to divide the portions equally, I had to divide two rectangles into smaller thirds.
3. Then I shaded in one-third of the array because that is what the problem asked for.
4. Now the shaded portion will be the product. There are $8^{1}/_{3}$ rectangles shaded.

Remember, the algorithm is to multiply numerator by numerator and denominator by denominator. When you have a fraction and a whole number, you have to change the whole number into the same unit of measurement, which means the whole number will be rewritten as the whole number over 1, and that the product will always be either an improper fraction or a mixed number. For example:

$$^{1}/_{3} \cdot {}^{25}/_{1} = {}^{25}/_{3} \longrightarrow 8^{1}/_{3}$$

Improper fraction   Mixed fraction

Note also that the test maker is assessing your understanding of how *fractions* that are multiplied by a *whole number* behave. Above, you are dividing the whole into *equal* portions based on the fractional part you are seeking, in this case $^{1}/_{3}$. What the test maker really wants to know is whether you understand that you must have 25 *equal portions* divided into 3 *equal sections* and that the shaded section is the product. The test maker also wants to know that you understand that whenever a fraction is multiplied by a whole number, the product will represent only a portion of the whole, and that portion will always be smaller than the whole.

Sometimes mixed fractions are multiplied. A simple algorithm would be to change the factors into improper fractions, then multiply both numerators and both denominators to get the product. For example:

$$5^{7}/_{8} \cdot 2^{4}/_{6}$$

$5^{7}/_{8} = {}^{47}/_{8}$ $2^{4}/_{6} = {}^{16}/_{6}$

5 times 8 equals 40 plus 7 equals $^{47}/_{8}$

2 times 6 equals 12 plus 4 equals $^{16}/_{6}$

$^{47}/_{8} \cdot {}^{16}/_{6} = {}^{752}/_{48} \longrightarrow \boxed{15^{32}/_{48} = 15^{2}/_{3}}$

Now let's look at the above expression from another perspective. The model below graphically shows what the expression $5^7/_8 \cdot 2^4/_6$ represents:

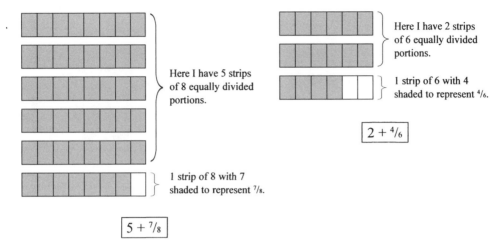

If we did not use the algorithm above to change the mixed fractions into improper fractions before multiplying, we could find the product according to the model above by multiplying each integer of the first expanded expression by the two integers of the second expanded expression and still have the same product. See below.

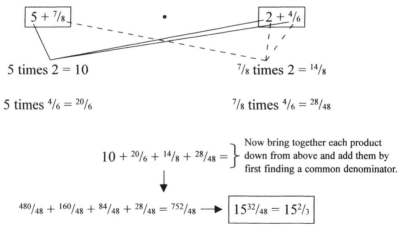

We just utilized what is called the *distributive property*. This property allows us to manipulate expressions so that they are more easily solved. It spreads out an expression to demonstrate how the integers are derived. An efficient way to find the product using the distributive property is to use a strategy called FOILing (F = First, O = Outside, I = Inside, L = Last). This is a technique used universally in higher math. For example, the model above can be numerically represented by $(^5/_1 + ^7/_8) \cdot (^2/_1 + ^4/_6)$. Notice that the whole numbers are rewritten as fractions but the expansion and process are the same as the model above.

Example using the FOIL method:

$$(7/8)(2/1) + (7/8)(4/6)$$

$$[5/1 + 7/8] \times [2/1 + 4/6]$$

$$(5/1)(2/1) + (5/1)(4/6)$$

$$14/8 + 28/48 + 10/1 + 20/6 = 752/48 \longrightarrow 15\,2/3 \;\Big\}\; \text{Each is the product from above.}$$

Dividing Fractions

Recall that division is repeated subtraction. Using whole numbers, we know that the quotient tells us how many times the divisor is subtracted from the dividend in its smallest units. For example:

$$30 \div 5 \;\Big\}\; 30 - \underline{5} = 25 - \underline{5} = 20 - \underline{5} = 15 - \underline{5} = 10 - \underline{5} = 5 - \underline{5} = 0$$

Notice that there are six 5s subtracted from 30. So $30 \div 5 = 6$.

The same concept applies to fractions. See below.

How many $1/8$'s are in $3/4$?

$$3/4 \div 1/8 \;\Big\}\; 3/4 - 1/8 = 6/8 - \underline{1/8} = 5/8 - \underline{1/8} = 4/8 - \underline{1/8} = 3/8 - \underline{1/8} = 2/8 - \underline{1/8}$$
$$= 1/8 - \underline{1/8} = 0$$

Notice that six $1/8$'s can be subtracted from $3/4$.

Although repeated subtraction will give a quotient, the process is laborious. This is why multiplying by the reciprocal is used. This process works because when you subtract a fraction, each time you have to find a common denominator before you can begin subtracting. This means that each portion has to be equal in size in order to subtract the same evenly. Watch what happens when we divide the two fractions above using cross-multiplication.

$$\frac{3}{4} \times \frac{1}{8} = \frac{24}{4} \longrightarrow 6 \qquad \text{When using this method, you have to remember which product is the numerator and which is the denominator.}$$

Notice the solution is still the same using cross-multiplication. This happens because we are really finding common denominators in their simplest form instead of going through the laborious process that follows.

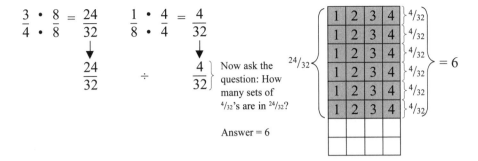

From the model above, an algorithm for dividing fractions was derived to simplify and clarify the process.

Process: Get the reciprocal of the divisor and change the operation to multiplication and then multiply across: numerator by numerator and denominator by denominator.

Finding the reciprocal means to flip the fraction ($^7/_8 \rightarrow {}^8/_7$), but not either fraction). It has to be the divisor. For example:

$^1/_4 \div {}^4/_6 = {}^1/_4 \cdot {}^6/_4 = {}^6/_{16} = {}^3/_8$ Notice that the second term (the divisor) has been flipped and the operation has changed from division to multiplication.

The divisor will always be on the right.

Why Does Multiplying by the Reciprocal Work?

Since division is the "opposite" of multiplication, and dividing by two is the same as multiplying by $^1/_2$, then division of fractions is most efficiently performed by multiplying the dividend by the reciprocal of the divisor.

All rational numbers (fractions) have multiplicative inverses except for zero. Zero has no reciprocal because it would either equal zero or be undefined. We use the multiplicative inverse when dividing fractions because that will not change the value of the expression; it only changes the way that it looks. It is a unique way to manipulate expressions so that you can easily solve them without changing the value versus using the cumbersome example at the beginning of this section. For example:

$$\frac{{}^9/_{12} \cdot {}^8/_3 = {}^{72}/_{36}}{{}^3/_8 \cdot {}^8/_3 = 1} = {}^{72}/_{36} = 2$$

Notice that we got the reciprocal of $^3/_8$, which is the same as getting the multiplicative inverse, and their product equals 1, and we also change the operation from division to multiplication. Simplification follows.

Here is the same example using the division of fractions rule:

$$9/12 \div 3/8 \rightarrow 9/12 \bullet 8/3 = 72/36 = 2$$

You can save yourself time and trouble by simplifying *before* calculating by using a process called cancelling. Note that this can be done with multiplication and division. When applying this strategy to division, remember that you must first get the reciprocal of the divisor before cancelling. Study this example:

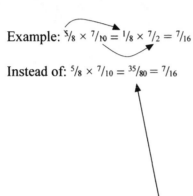

Example: $5/8 \times 7/10 = 1/8 \times 7/2 = 7/16$

Instead of: $5/8 \times 7/10 = 35/80 = 7/16$

1. Look for common factors diagonally across the operation sign. Notice that 5 and 10 both have a common factor of five.
2. Now divide both numbers by the common factor. $5 \div 5 = 1$ and $10 \div 5 = 2$. Rewrite the expression with the simplified solutions.
3. Now look for common factors diagonally across the other two factors to see if they have any common factors. Here they do not because 7 and 8 have no common factors. However, if they did, you would go through the same process and then multiply numerator by numerator and denominator by denominator. All you are doing is simplifying before multiplying, otherwise you will have to simplify at the end.

Division is done in a similar way; just remember to invert the second number before multiplying.

Example: $6/7 \div 3/5 = 6/7 \times 5/3 = 2/7 \times 5/1 = 10/7 = 1\ 3/7$

Example: $1/3 \div 3/4 = 1/3 \times 4/3 = 4/9$

Now let's look at some word problems using these two operations.

Example: John and Bill share $64. If John's share is $1/4$ of $64, how much money does Bill receive?

Answer: $1 - 1/4 = 3/4$ (This is Bill's share.)
$64/1 \times 3/4 = 192/4 = \$48$

Example: A can of tomatoes holds $2\frac{1}{2}$ cups and is used in a recipe to serve six people. If Dan wants to serve two people, how many cups of tomatoes must he use?

Answer: Two people require $2/6 = 1/3$ of the amount of tomatoes.
$2\frac{1}{2} \times 1/3 = 5/2 \times 1/3 = 5/6$ cups of tomatoes

Note: You always need to convert a mixed number into an improper fraction before calculating: $2\frac{1}{2} = 5/2$ $(2 \times 2) + 1 = 5$.

Example: How many quarter-inch pieces can be cut from a board whose length in inches is:

A. 15
B. $18\frac{1}{2}$
C. $25\frac{3}{4}$

Answers:

A. $15 \div 1/4 = 15/1 \div 1/4 = 15/1 \times 4/1 = 60$ pieces
B. $18\frac{1}{2} \div 1/4 = 37/2 \div 1/4 = 37/2 \times 4/1 = 37/1 \times 2/1 = 74$ pieces
C. $25\frac{3}{4} \div 1/4 = 103/4 \div 1/4 = 103/4 \times 4/1 = 103/1 \times 1/1 = 103$ pieces

Example: How many poles, each $1\frac{1}{2}$ feet long, can be cut from a longer pole whose length in feet is:

A. $22\frac{1}{2}$
B. 30
C. $37\frac{1}{2}$

Answers:

A. $1\frac{1}{2} = 3/2$
$22\frac{1}{2} \div 3/2 = 45/2 \div 3/2 = 45/2 \times 2/3 = 15/1 \times 1/1 = 15$ pieces
B. $30 \div 3/2 = 30 \div 3/2 = 30 \times 2/3 = 10/1 \times 2/1 = 20$ pieces
C. $37\frac{1}{2} \div 3/2 = 75/2 \div 3/2 = 75/2 \times 2/3 = 25/1 \times 1/1 = 25$ pieces

Example: A pie crust requires $3\frac{3}{8}$ cups of flour, $1\frac{1}{4}$ teaspoons of salt, and $5/6$ cups of water. How much of each ingredient is needed for eight pie crusts?

Answers:

Flour: $3^3/_8 \times {}^8/_1 = {}^{27}/_8 \times {}^8/_1 = {}^{27}/_1 \times {}^1/_1 = 27$ cups of flour

Salt: $1^1/_4 \times {}^8/_1 = {}^5/_4 \times {}^8/_1 = {}^5/_1 \times {}^2/_1 = 10$ teaspoons of salt

Water: ${}^5/_6 \times {}^8/_1 = {}^5/_3 \times {}^4/_1 = {}^{20}/_3 = 6^2/_3$ cups of water

Example: A metal rod weighs $2^1/_3$ pounds per foot. How many pounds does a rod weigh if its length is:

A. 9 feet

B. 12 feet

C. $5^1/_2$ feet

Answers:

A. Note that $2^1/_3 = {}^7/_3$
   ${}^9/_1 \times {}^7/_3 = {}^3/_1 \times {}^7/_1 = 21$ pounds

B. ${}^{12}/_1 \times {}^7/_3 = {}^4/_1 \times {}^7/_1 = 28$ pounds

C. $5^1/_2 \times {}^7/_3 = {}^{11}/_2 \times {}^7/_3 = {}^{77}/_6 = 12^5/_6$ pounds

END OF PRAXIS I, 0014, 0511, and 0146

These tests require the same skills but at a higher level of complexity. Here are some worked problems that illustrate this type of question:

Example: $1/_4$ is the mean of $1/_5$ and what number?

Answer: Remember the formula, Mean = $\Sigma X/N$

$(1/_5 + x)/2 = 1/_4$

$1/_5 + x = 2(1/_4)$

$1/_5 + x = 1/_2$

$10(1/_5) + 10x = 10(1/_2)$

$2 + 10x = 5$

$10x = 5 - 2$

$10x = 3$

$x = {}^3/_{10}$

Check: Does the mean of $3/_{10}$ and $1/_5 = 1/_4$?

$3/_{10} + 1/_5 = 3/_{10} + 2/_{10} = 5/_{10} = 1/_2$

$1/_2 \cdot 1/_2 = 1/_4$

The answer is correct.

Example: What is the reciprocal of $3^7/_8$?

Answer: The product of a number and its reciprocal is 1.
Start by converting to an improper fraction: $3^7/_8 = {}^{31}/_8$.
$^8/_{31}$ is the reciprocal because $^{31}/_8 \cdot {}^8/_{31} = 1$.

Example: What is the reciprocal of $3^3/_4$?

Answer: Since $3^3/_4 = {}^{15}/_4$, the reciprocal is $^4/_{15}$.

Example: Katie is buying curtain fabric for three windows, each 40 inches wide. She needs to buy fabric that is $2^2/_3$ times as wide as the windows. How much fabric should she purchase?

Answer: The total width of the windows is $3 \cdot 40$, or 120 inches.
Now multiply that by $2^2/_3$ ($^8/_3$).
$120 \cdot {}^8/_3 = 40 \cdot 8 = 320$ inches (or $320/12 = 26^2/_3$ feet or 26 feet, 8 inches)

Example: George and his family ordered a pizza and ate $^2/_3$ of it. The next day his sister took half of what was left for lunch. What portion of the original pizza did she have for lunch?

Answer: The family ate $^2/_3$ for dinner so $^1/_3$ was left. George's sister took $^1/_3 \cdot {}^1/_2 = {}^1/_6$ of the original pizza for her lunch.

Example: Freddie needs to read 14 pages for science, 26 for civics, 12 for English, and 28 for history. He has already read $^1/_6$ of his assignments. How many pages has he read?

Answer: He was assigned to read $14 + 26 + 12 + 28 = 80$ pages.
$^1/_6$ of $80 = {}^1/_6 \cdot 80 = {}^{80}/_6 = 13^2/_6 = 13^1/_3$ pages

Example: A full spaghetti recipe calls for $3^1/_2$ pounds of noodles. How many pounds would be needed to make $^1/_3$ of the recipe?

Answer: $3^1/_2 \cdot {}^1/_3 = {}^7/_2 \cdot {}^1/_3 = {}^7/_6 = 1^1/_6$ pounds

Example: Cindy lives $5^1/_3$ miles from school. She has to walk $^1/_4$ of that distance to reach her bus stop. How far is the bus stop from her house?

Answer: Simply multiply the total distance by $^1/_4$ to find the distance to the bus stop.
$5^1/_3 \cdot {}^1/_4 = {}^{16}/_3 \cdot {}^1/_4 = {}^{16}/_{12} = {}^4/_3 = 1^1/_3$ miles to the bus stop.

Example: In August, a taxi company had $1/6$ of the taxis in for maintenance. Repairs were also being done on $1/8$ of the vehicles. If those two categories comprised 28 vehicles, how many taxis did the company have?

Answer:

$(1/6 \cdot x/1) + (1/8 \cdot x/1) = 28$

$x/6 + x/8 = 28$

$4x/24 + 3x/24 = 28$

$7x/24 = 28$

$x = 28 \cdot 24/7$

$x = 4 \cdot 24 = 96$ taxis in all

Example: If $3/11$ of a number is 22, what is $6/11$ of that number?

Short cut: You could simply say $6/11$ is twice $3/11$, so the answer will be twice 22, or 44. Thinking this way can save a significant number of minutes on a long timed test. The time saved can be profitably invested on other questions where no shortcuts are available.

Answer:

$(3x)/11 = 22$

$3x = 22(11)$

$3x = 242$

$x = 242/3$

Then $242/3 \cdot 6/11 =$

$22/1 \cdot 2/1 = 44$

Example: $5/8$ is equal to $15/7$ of what number?

Answer:

$15/7 \cdot x/1 = 5/8$

$x = 5/8 \cdot 7/15$

$x = 7/24$

Check: Does $7/24 \cdot 15/7 = 5/8$?

$15/24 = 5/8$

$5/8 = 5/8$

Yes, the answer is correct.

Example: Billy won some goldfish at the state fair. During the first week $1/5$ of them died, and during the second week $3/8$ of those still alive at the end of the first week died. What fraction of the original goldfish were still alive after two weeks?

Answer: It is more helpful to think of the fractional part alive than dead. So at the end of the first week $4/5$ will still be alive ($5/5 - 1/5$). That makes the final calculation very straightforward:

$4/5 \cdot 5/8 = 1/1 \cdot 1/2 = 1/2$

Example: What is the value of the following product?

$5/5 \cdot 5/10 \cdot 5/15 \cdot 5/20 \cdot 5/25$

Answer: This reduces to:

$1 \cdot 1/2 \cdot 1/3 \cdot 1/4 \cdot 1/5 = 1/120$

12

Fractions to Decimals to Percents

INTRODUCTION

A common mistake made by many students is attending to a multitude of details instead of gaining an understanding of "the big picture." The format suggested in this chapter is an attempt to present one of these understandings to replace an overattention to details.

A further benefit of seeing the big picture is that it is much easier to remember than a multitude of small procedures. And to the degree that this is so, test scores may be increased as well.

RELEVANT CONCEPTS FOR ALL TESTS

Many test takers fail to earn all the points that they deserve because they do not fully understand the relationships among fractions, decimals, and percents. They mistakenly think that these three concepts are distinct and separate rather than related. The following diagram (and the process below) illustrates clearly the relationships, and it can be reproduced by students on a test for instant consultation.

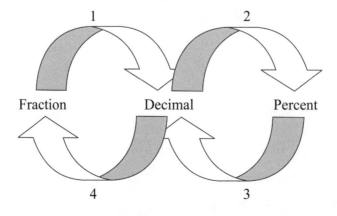

1. *Conversion of fractions to decimals*: Divide the numerator by the denominator. For example, ⁵/₈ is 5 divided by 8, or $8\overline{)5}$ = .625. Always divide the "top" number (numerator) by the "bottom" number (denominator).

2. *Conversion of decimals to percents*: Move the decimal point two places to the right and add a percent symbol. The mathematical operation is actually multiplication by 100. For example, .625 = 62.5%.

3. *Conversion of percents to decimals*: Move the decimal point two places to the left and drop the percent symbol. The mathematical operation is division by 100. For example, 62.5% = .625.

4. *Conversion of decimals to fractions*: Say correctly, using the place values like tenths, hundredths, and thousandths. Then write as a fraction from the words. Finally, reduce if possible. For example, .625 is six hundred and twenty-five thousandths. Next, it is written as ⁶²⁵/₁₀₀₀. Numerator and denominator can be divided by 5, 25, and 125. Keep reducing until the fraction is in lowest terms, in this case ⁵/₈.

THE NATURE OF THE RELATIONSHIP AMONG THESE THREE CONCEPTS

Changing Percents to Fractions

A percent is a unique kind of fraction or part of a whole. A fraction is composed of two whole numbers. It has three parts: a numerator (the top number), a denominator (the bottom number), and the bar that is between the numbers (which is a symbol that means to divide). For example:

$$\frac{3}{8} \quad \frac{\text{Numerator}}{\text{Denominator}} \longleftarrow \text{Means "to divide"}$$

What makes a percent unique is that its denominator will always be 100. For example, 5% can be written as ⁵/₁₀₀. The word "percent" literally means "out of 100." Thus 5% means 5 parts out of 100.

Another more complicated example is 16²/₃%. The percent sign beside this expression says that this whole expression is over 100:

$$\frac{16^{2}/_{3}}{100}$$

To change a percent into a fraction, simply remove the percent symbol and put the expression over 100, and then simplify the expression. See the following procedure.

Simple percent conversion

$5\% \rightarrow \dfrac{5}{100} \dfrac{\div 5}{\div 5} = \dfrac{1}{20}$

1. Change 5% to a fraction by removing the percent symbol and placing the 5 over 100.
2. Simplify the fraction to its lowest terms by dividing both numerator and denominator by 5.

Complex percent conversion

$16\tfrac{2}{3}\% \rightarrow \dfrac{16^{2}/_{3}}{100} \rightarrow \dfrac{^{50}/_{3}}{100} \rightarrow \dfrac{50}{3} \div \dfrac{100}{1}$

$\dfrac{\overset{1}{\cancel{50}}}{3} \cdot \dfrac{1}{\underset{2}{\cancel{100}}} = \dfrac{1}{6}$

1. Change $16^{2}/_{3}\%$ to a fraction by removing the percent symbol and writing $16^{2}/_{3}$ over 100.
2. Rewrite the numerator as an improper fraction over 100.
3. Remember that the bar indicates division, so divide these two fractions.
4. Remember that when dividing fractions, you get the reciprocal of the divisor and change the operation from division to multiplication.
5. Simplify the fraction to its lowest terms, either by cancelling first or multiplying and then simplifying.

Changing Fractions to Decimals

To change a fraction into a decimal, always divide the numerator by the denominator.

$5 \div 8$ or $^{5}/_{8}$ means 5 divided by 8, which means $8\overline{)5} = .625$

All fractions with whole numbers in both numerator and denominator will either terminate (end with a remainder of zero) or repeat digits in some order. If the denominator has *only* prime factors of 2 or 5 or the combination of 2 and 5, then the decimal will terminate (have a remainder of zero); otherwise the quotient will repeat. Using the example above, 8 is in the denominator or divisor. It can be rewritten as 2^3, which means $2 \cdot 2 \cdot 2$. Notice that the prime factors are only twos, thus 8 will divide into 5 evenly in decimal form, which is .625. This same idea will result with a denominator that has only 5 as a prime factor. For example, $^{21}/_{25} = 25\overline{)21} =$ 21 divided by 25. Twenty-five can be rewritten as 5^2, which means $5 \cdot 5$. From this

Fractions to Decimals to Percents 83

we know that our quotient will terminate because the divisor has only prime factors of 5. The quotient equals .84.

Following is one more example using both prime factors of 2 and 5. This should give you a clue that the divisor has to be a multiple of 10 because 2 times 5 equals 10. Let's use $40\overline{)63} = 1.575$. Forty can be rewritten as $2^3 \cdot 5^1$. Notice that it has only 2 and 5 as prime factors. So I know that my quotient should terminate with three digits to the right of the decimal because 2 and 5 are prime factors. This characteristic is important in understanding the relationship between decimals, fractions, and percents.

What Is the Underlying Relationship between These Three Concepts?

The Power of 10. Decimals are powers of 10 because .1 is one tenth, or $^1/_{10}$; .01 is one hundredth, or $^1/_{100}$; .001 is one thousandth, or $^1/_{1000}$, and so on. As noted above, their equivalents can be rewritten as fractions of the power of 10. And since percent means "out of 100," they are also related to the powers of 10 because 100 is a power of 10 rewritten as 10^2, and by extension 2 and 5 are prime factors of 10 that will always terminate when used as a denominator. This phenomenon is modeled below.

Terminating Decimal Example Using the Prime Factor 5

$5\overline{)63},\ 25\overline{)23},\ 125\overline{)63}$ Any of these divisors will terminate because they are powers of 5, which means they have prime factors of only 5. For example:

$$\begin{array}{c} 25 \\ / \quad \backslash \\ 5 \quad\quad 5 = 5^2 \end{array} \qquad\qquad \begin{array}{c} 125 \\ / \quad \backslash \\ 5 \quad\quad 25 \\ \quad\quad / \backslash \\ \quad\quad 5 \ \ 5 = 5^3,\ \text{etc.} \end{array}$$

Let's use $25\overline{)23}$.

1. $25\overline{)23.0}$ 25 will not divide into 23, so show the decimal at the end of the number and add a zero (a zero because decimals are a power of 10).

2. $\begin{array}{r} .9 \\ 25\overline{)23.0} \\ \underline{225} \\ 5 \end{array}$ Now 25 will divide into 230 nine times. Nine times 25 equals 225.

```
       .92
3. 25)23.00
       225↓
        50
        50
         0
```

Subtract 225 from 230 to get the remainder of 5. Add another zero to the dividend and bring it down beside the 5 to make it a 50. Now 25 will divide into 50 two times with no remainder. The quotient is .92. Also notice that the power determines the number of decimal places (5^2—two is the exponent so the quotient will extend two places after the decimal point and then terminate).

Terminating Decimal Example Using the Prime Factor 2

Any of these divisors will terminate because they are powers of 2, which means they have prime factors of only 2. For example:

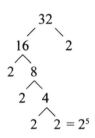

Let's use $32\overline{)23}$.

```
       .7
1. 32)23.0
      224
        6
```

32 will not divide into 23, so show the decimal after the 3 and add a zero (a zero because decimals are a power of 10).

```
       .71
2. 32)23.00
      224↓
       60
       32
       28
```

Now 32 will divide into 230 seven times. Seven times 32 equals 224. Subtract 224 from 230 to get the remainder of 6.

32 will not divide into 6, so add another zero to the dividend and bring it down beside the 6 to make it a 60. Now 32 will divide into 60 once. One times 32 equals 32. Now subtract 32 from 60. This gives a remainder of 28.

```
         .718
3. 32) 23.00
       224
        60
        32
       280
       256
        24
```
32 will not divide into 28 so add another zero and bring it down beside 28, making it 280. 32 will divide into 280 eight times. Eight times 32 equals 256. Subtracting 256 from 280 equals 24. This will go two more times before it will terminate. I know this because 2^5 is the prime factorization of 32. The exponent is 5, so this means the decimal will have five places. I already have three places, so I will need two more before the quotient finally terminates at .71875.

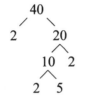   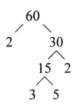

Notice that with 40, the prime factorization is $2^3 \cdot 5$, so it fits the rule, whereas 60 does not. Its prime factorization is $2^2 \cdot 5 \cdot 3$. 60 has an additional prime factor (3). It doesn't have prime factors of only 2 and 5.

Changing a Percent to a Fraction and then to a Decimal, and Vice Versa

Remember that the underlying relationship between a percent, a fraction, and a decimal is the power of 10, and that a decimal can be rewritten as a fraction using the power of 10. We discussed how to change a fraction into a decimal. Now let's use the relationship of the power of 10 to show how a percent is nothing more than a fraction that can easily be converted into a decimal.

A percent is a number out of 100, which could also be called a special ratio. A ratio is a comparison of two quantities by division, which means that a percent can be rewritten as a fraction.

For example, 6% = $^6/_{100}$. We can do this because the actual percent symbol (%) is equivalent to saying "hundredths" (.01) and therefore can be expressed as a fraction with 100 as the denominator. Once the percent is rewritten as a fraction, the percent sign is no longer needed. Then we can convert the fraction into a decimal by dividing the denominator into the numerator. See the process on the following page.

$$6\% = {}^6/_{100} = 100\overline{)6.00}$$

Division:
- .06
- 100)6.00
- 600
- 0

1. 100 will not divide into 6, so put a decimal after the 6 and add a zero, and then place a decimal in the quotient aligned with the decimal in the dividend.
2. Ask, "How many times will 100 go into 60?" It will not, so put a zero after the decimal in the quotient and add another zero in the dividend.
3. Now ask, "How many times will 100 go into 600?" It will go 6 times with no remainder.

We have succeeded in changing a percent to a fraction to a decimal. Notice that the quotient is six hundredths, which is a form of the power of 10.

Now let's undo the process and change the quotient .06 (six hundredths) back to its original form (6%). When a decimal is written as the power of 10, it can easily be converted back to fractional form by simply putting the number over 100 (.06 = $^6/_{100}$).

Now notice the pattern below and how the power of 10 behaves with a decimal.

$.6 = {}^6/_{10}$	$.06 = {}^6/_{100}$	$.006 = {}^6/_{1000}$
.6 has one place to the right of the decimal and the denominator has one zero.	.06 has two places to the right of the decimal and the denominator has two zeroes.	.006 has three places to the right of the decimal and the denominator has three zeroes.

Notice that the number of places after the decimal determines how many zeroes are placed after the one in the denominator because of its relation to the power of 10. Using this idea, we can easily change a decimal to a percent and vice versa.

To change a *decimal* to a *percent*, the decimal is really multiplied by 100 because percent means "out of 100." For example:

$.6 \cdot 100 = 60.0\%$ $.06 \cdot 100 = 6.00\%$ $.006 \cdot 100 = .6\%$

Notice that each time the decimal is multiplied by 100, the decimal point moves to the right two times. So multiplying by 100 is really not necessary when converting decimals to percents because all that is needed is to remember to move the decimal two places to the right and to place a percent symbol beside the number. If the number has trailing zeroes after the decimal point, the zeroes and the decimal point can be omitted, otherwise keep the decimal point as part of the solution. For example, .532 equals 53.2% (the decimal point remains but has moved two places to the right). This idea follows the same reasoning as the above fractional example.

If I want to change a *percent* back to a *decimal*, divide by 100 or simply move the decimal two places to the left and drop the percent symbol.

Fractions to Decimals to Percents 87

```
    .06
100)6.00    or   6%  =  .06
    600
      0
```

Here a decimal is always understood to be at the right of a number. We moved it two places to the left and dropped the percent symbol.

Test makers may consciously ask a direct question related to the above format.

Example: 37.5% equals what fraction?

Answer: From the diagram at the beginning of this chapter, we see that two steps are involved.

Step one: Convert the *percent* to a *decimal*.
37.5% = .375

Step two: Convert the *percent* to a *fraction*.
.375 = three hundred and seventy-five thousandths

Written as a fraction, that becomes $^{375}/_{1000}$. This will reduce to $^{75}/_{200}$ = $^{15}/_{40}$ = $^{3}/_{8}$, which is the answer.

In other variations of these concepts, the data may come from a table or graph but the computations are exactly the same.

Example: Suppose you own a freight company and you have budgeted costs for shipping based on weight and distance. The table below represents the cost as a percentage of weight.

You have an order to ship a product to Tennessee with the weight of 200 pounds. How much will you charge?

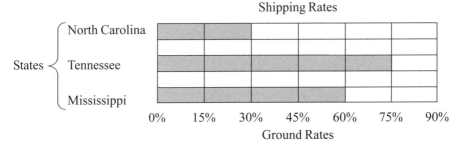

Answer:

1. Look at the table to locate the rate for Tennessee, which is 75%.

2. Change the percent to a decimal (.75).

3. Multiply .75 by 200 pounds.

4. .75 • 200 pounds = $150.00.

Another question that can be solved easily with this subject unifier might be the following:

Example: Arrange the following fractions in ascending order (smallest to largest).

$3/8 \quad 1/4 \quad 2/7 \quad 5/14 \quad 5/8$

Answer: Though it would be possible to answer by careful inspection, a more efficient approach might be to convert each fraction to a decimal and place the conversion underneath the fraction. That would give the following:

$3/8 \quad 1/4 \quad 2/7 \quad 5/14 \quad 5/8$
.375 .250 .286 .357 .625

Then decide ascending order by using this process:

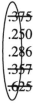

1. Circle the largest numbers (tenths) and decide which is the largest *tenth* from the group of numbers. The number .625 is the largest and goes at the bottom of the chart because the example is asking for ascending order (least to greatest).

2. You have two decimals that could be the next largest (.3$\underline{7}$5 and .3$\underline{5}$7). Look at the next place (hundredths) to decide which is larger. Notice that we underlined the 7 and 5. Seven is larger than 5, so .375 will be next, followed by .357. Now place these two numbers from the bottom going upward.

3. Mark through the ones that are done to see what numbers are left.

4. Circle the next largest numbers (tenths) and eliminate the ones that are slashed. We have two possible numbers (.2$\underline{5}$0 and .2$\underline{8}$6). Look at the next place value (hundredths). Eight is larger than 5, so .286 will be next then .250. Place these two numbers in the chart accordingly.

5. Insert fractions beside the decimals to prevent mistakes. Now it will be easy to identify the correct solution.

Smallest to largest:	
.250	1/4
.286	2/7
.357	5/14
.375	3/8
.625	5/8

Another approach using the same unifier is *Percent of Change*.

Sometimes percents are useful to show change. The new amount will be compared as an increase or decrease to the original number in the form of a percent. Because the process is exactly the same whether you are finding percent increase or decrease, it is important that you are able to identify whether the change is an increase or a decrease.

The following illustrates a percent increase:

Example: A school experiences an *increase* in enrollment from 571 students to 623 students. What is the *percent increase*?

Answer: Express as

$$\frac{\text{increase}}{\text{original enrollment}} \longleftarrow \frac{\text{Numerator}}{\text{Denominator}} \longleftarrow \text{Means "to divide"}$$

$$\frac{(623 - 571)}{571} = \frac{52}{71} = .091 = 9.1\% \text{ increase}$$

Here is the process to follow:

1. Always subtract the smaller number (571) from the larger number (623).
2. Divide the difference by the original number (571).
3. Change the fraction to a decimal by dividing the denominator into the numerator $(571\overline{)52})$.
4. Now change the decimal to a percent by moving the decimal two places to the right. Place a percent sign beside the number and label the solution as an increase.

$571\overline{)52.}^{.091}$

$.091 \rightarrow 9.1\%$

Note: The 0 in front of the 9 can be omitted because it has no value when there is no number to its left.

The following illustrates a percent decrease:

Example: A department store reduced sweaters in April from $39.99 to $24.99. What was the *percent decrease*?

Answer: Express as

$$\frac{\text{decrease}}{\text{original price}} \longleftarrow \frac{\text{Numerator}}{\text{Denominator}} \longleftarrow \text{Means "to divide"}$$

$$\frac{(39.99 - 24.99)}{39.99} = \frac{15}{39.99} \approx .375 \approx 37.5\% \text{ increase}$$

Note: Percent decrease is the same process as percent increase except for labeling:

1. The original number is always in the denominator.
2. The new number is always subtracted from the original number.
3. Label the solution as 37.5% decrease.

Example: What are some equivalents to $^{13}/_{25}$?

Answer: $13 \div 25 = .52$ (Decimal Equivalent), $.52 = 52\%$ (Percent Equivalent), $^{26}/_{50}$ (Functional Equivalent, one of many)

Example: Ten students from the 281-member student senior class at Lexington High School received full college scholarships. What percent of the senior class was this?

Answer: $^{10}/_{281} = 10 \div 281 = .036 = 3.6\%$ of the seniors.

Example: In order to pass a certain exam, firefighters must answer 75% of the questions correctly. If there were 85 questions on the exam, how many must be answered correctly?

Answer: 75% of $85 = .75 \times 85 = 63.75$.

Note: Since we are looking at whole questions, any fractional part must be rounded up. This gives 64 correct answers.

Example: Henry earned a $5^3/_4\%$ pay raise. If his salary was $28,300 before the raise, how much was his salary after the raise?

Method 1 (2-step)	**Method 2 (1-step)**
1. Convert the percent to a decimal (.0575).	1. Since this is an increase, *add* 100% to the raise (100% + 5.75% = 105.75%).
2. Multiply the decimal by the old salary (.0575 • $28,300 = $1,627.25).	2. Change the percent to a decimal by moving the decimal two places to the left (1.0575).
3. Add the old salary to the pay raise ($28,300 + $1,627.25 = $29,927.25).	3. Now multiply 1.0575 by $28,300 (1.0575 • $28,300 = $29,927.25).

Method 3 (1-step)

1. Treat the original salary as 1 because it is considered one whole.
2. Change the percent to a decimal (.0575).
3. Now add the 1 and .0575 together (1 + .0575 = 1.0575).
4. Now multiply 1.0575 by $28,300 (1.0575 • $28,300 = $29,927.25).

Methods 2 and 3 have the distinct advantage of yielding the final answer directly in one calculation. Some students will use the two-step approach and look down at the answers after just the first step. Since test makers often use a partial answer as one of the incorrect choices, these students may miss the question for that reason only. A one-step approach avoids this possibility.

Example: Booster club members voted on how to donate money raised to each athletic program in their school. The pie chart shows what they decided. $165.00 was donated to the football program. How much money did the booster club members raise? Then show how much money was given to each of the other sports.

Process for the first part of the example:

1. Change 40% to a decimal by moving the decimal point two places to the right and dropping the percent sign (.40).
2. Divide $165.00 by .40
($165.00 = .40 • X)
($165.00 ÷ .40 = X)
($412.50 = X)

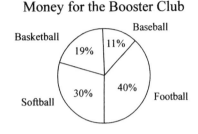

This means $412.50 is the total raised by the booster club members.

Process for the second part of the example:

Since I know the total money raised, I can easily find how much money was given to the rest of the programs.

Softball: Change 30% to a decimal, then multiply (.30 • $412.50 = $123.75).

Basketball: Change 19% to a decimal, then multiply (.19 • $412.50 = $78.38).

Baseball: Change 11% to a decimal, then multiply (.11 • $412.50 = $45.38).

END OF PRAXIS I, 0014, 0511, AND 0146

The kinds of problems that could appear on other tests can be complicated. Often there is a combination of skills tested that goes beyond the conversion of fractions to decimals to percents. They may involve ratios, proportions, and even geometry.

Here are several sample problems of the correct level of difficulty that relate to conversions.

Example: If a person has an income of $100,000, what percent of his income does he pay in federal income tax if the tax rate is:

15% of the first $30,000

28% of the next $30,000

31% of income above $60,000

Answers:

If the first $30,000 is taxed at 15%, then that amount is $30,000 • .15 = $4,500.

If the next $30,000 is taxed at 28%, then that amount is $30,000 • .28 = $8,400.

If the last $40,000 is taxed at 31%, then that amount is $40,000 • .31 = $12,400.

The sum of the three amounts is $25,300 (4,500 + 8,400 + 12,400).

The fractional part of the total income is $25,300/100,000$.

The decimal equivalent is .253.

The percent is 25.3%.

Example: Jar B has 20% more marbles than jar A. What percent of the marbles in jar B has to be moved to jar A so the numbers will be equal?

Answer: If you assume that jar A has 200 marbles, then jar B would have 240 marbles (200 • 1.2). Equality would be achieved with the transfer of 20 marbles from jar B to jar A. You could assume many numbers (100, 300, 400, etc.) and the answer is always the same. This choice is not critical.

Fractional part of jar B moved: $20/240 = 2/24 = 1/12$.

Decimal equivalent (1 divided by 12): .083.

Percent moved: 8.3%.

Example: In a large jar full of jelly beans, 30% of them are red; 40% of the red jelly beans are cherry and 25% are raspberry. What percent of all the jelly beans are either cherry or raspberry?

Short cut: Just write the answer like this:

Cherry + Raspberry

(.3)(.4) + (.3)(.25) = .12 + .075 = .195 = 19.5%

Longer procedure for answer: Assume the jar holds 1,000 jelly beans (you could use other numbers and the answer would not change).

300 would be red (1,000 • .3).

40% of the red jelly beans will equal 120 cherry (300 • .4).

25% of the red jelly beans will equal 75 raspberry (300 • .25 = 75).

Now add the 120 cherry plus 75 raspberry (120 + 75 = 195).

195 of 1,000 = 195/1,000 = .195 = 19.5%.

Be careful. They want you to say 65% (40% + 25%), but that is clearly incorrect!

Example: On a test consisting of 80 questions, Marie answered 75% of the first 60 questions correctly. What percent of the last 20 questions did she need to answer correctly for her grade on the whole exam to be 80%?

Answer: Her number correct on the first 60 would be 60 • .75 = 45. She needs to have 80% correct of 80 questions, or 80 • .8 = 64 correct. The difference is 64 − 45, or 19 more correct on the last 20 questions.

Fractional representation: $^{19}/_{20}$.

Decimal equivalent: .95.

Percent equivalent: 95%.

Example: A medical insurance policy pays on the following scale:

Dollar Limits	**Percent Paid**
First $20,000	80%
Next $40,000	60%
Next $40,000	40%

If the medical bill is $92,000, how much will this policy pay?

Answer:

1. All the percents must be converted to decimals. In order they are .8, .6, and .4.
2. The bill must be broken down into appropriate segments by policy stipulations: $20,000, $40,000, and $32,000 (total of $92,000).
3. The calculation may be performed as follows:

    $20,000 • .8 + $40,000 • .6 + 32,000 • .4

    $16,000 + $24,000 + $12,800

    $52,800 = payout

13

Ratios and Proportions

INTRODUCTION

Information in a math question is often given in terms of something else rather than itself, as in most equations. This is the case with ratios and proportions. The *verbal relationship* of a proportion is:

$$A:B::C:D$$

That is, as "B" is related to "A," so "D" is related to "C." This also means that "A" and "C" are related, as are "B" and "D."

RELEVANT CONCEPTS FOR ALL TESTS

Proportions can also be thought of as fractions:

$$A/B = C/D$$

Which gives the following approach to a solution:

$$A \times D = C \times B$$

Often three of the four elements are given and the solution is to solve for the fourth number.

Ratios are really just simple fractions, and a solution is found quite directly.

Example: If $2/3$ of the workers in an office are nonsmokers, what is the ratio of smokers to nonsmokers?

Answer: They are only asking for you to create a fraction with smokers as the numerator and nonsmokers as the denominator. Like this:

$$\frac{\text{Smokers}}{\text{Nonsmokers}}$$

Of every three employees, one smokes and two do not. This gives the ratio of 1:2 or $1/2$ or 1 to 2. All are correct.

Example: If 80% of the applicants to a program were rejected, what is the ratio of the number accepted to the number rejected?

Answer: If 80% are rejected, then 20% are accepted. So for a typical 10 applicants, 8 are rejected and 2 are accepted. For the ratio of accepted/rejected, this gives $2/8$ or $1/4$ (or 1 to 4 or 1:4).

A question may include both numbers and symbols.

Example: What is the ratio of the circumference of a circle to its radius?

Answer: Start with the formula $c = 2\pi r$. The question seeks to know the value of c/r. Therefore:

$c = 2\pi r$

$c/r = 2\pi r/r$

$c/r = 2\pi$

Example: If x is a positive number and $x/3 = 12/x$, what is the value of x?

Answer: Apply cross-multiplication from above:

$x/3 = 12/x$

$x \times x = 3 \times 12$

$x^2 = 36$

$x = \sqrt{36}$

$x = 6$

Since it is always good to check your work, then does $6/3 = 12/6$?

$6 \times 6 = 3 \times 12$

$36 = 36$, so the answer is correct.

Example: At Lakeview High School the ratio of the number of students taking Spanish to the number of students taking French is 7:2. If 140 students are taking French, how many students are taking Spanish?

Answer: The two ratios are French:Spanish. So:

7:2::x:140 or

7/2 = x/140

2x = 7 × 140

2x = 980

x = 980/2

x = 490, which is the number taking Spanish.

To check, does 7/2 = 490/140 ?

7/2 = 49/14

7 × 14 = 2 × 49

98 = 98, so the answer is correct.

It is possible to have more than two components in proportions.

Example: Three partners in a business share profits in the ratio of 2:3:5. Find the share of each partner if the profits are $20,400.

Answer: Note that if the ratios are 2:3:5, then the first gets 2:10, the second gets 3:10, and the third gets 5:10 (2 + 3 + 5 = 10, or the total number of shares).

The first partner, who receives a share of 2, would get:

    2/10 = x/20,400

    10x = 2 × 20,400

    10x = 40,800

    x = 4,080

The second partner, who receives a share of 3, would get:

    3/10 = x/20,400

    10x = 3 × 20,400

    10x = 61,200

    x = 6,120

The third partner, who receives a share of 5, would get:

    5/10 = x/20,400

    10x = 5 × 20,400

    10x = 102,000

    x = 10,200

Example: In Filipe's class there are 4 women to 3 men. If there are 28 students, how many of them are men?

Answer: Set up the proportion like this:

3/7 = x/28

7x = 3 × 28

x = (3 × 28)/7

x = 3 × 4 = 12 men

END OF PRAXIS I, 0014, 0511, AND 0146

At this level the problems are similar but a bit more complex.

Example: In the diagram below, b:a::7:2. What is the value of b – a?

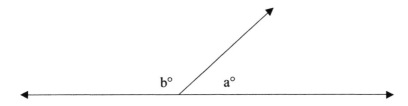

Answer: Since the angles are in the ratio of 7:2, then there are 9 parts to the sum of the two angles (180°) and:

7/9 = b/180

9b = 7 × 180

9b = 1,260

b = 1,260/9

b = 140°

And:

2/9 = a/180

9a = 2 × 180

9a = 360

a = 360/9

a = 40°

The two angles are 140° and 40° (which makes sense because the sum is 180°).

Now, finally the question may be answered.

b – a = 140 – 40

b – a = 100

Example: If a:b::3:5 and a:c::5:7 what is the value of b:c?

Answer:

$a/b = 3/5$

$a = 3b/5$

$a/c = 5/7$

$a = 5c/7$

And thus:

$3b/5 = 5c/7$

$3b/5c = 5/7$

$(3b/5c)5 = 5/7(5)$

$3b/c = 25/7$

$3b/c(1/3) = 25/7(1/3)$

$b/c = 25/21$ or 25:21 (or 25 to 21)

Example: If $a/b = 2/3$ and $b/c = 4/5$, what is the value of a/c?

Answer: Since $a/b \times b/c = a/c$, then the answer is simply the product of the two fractions.

$^2/_3 \times ^4/_5 = ^8/_{15}$ (or 8:15 or 8 to 15)

Example: A club had 3 boys and 5 girls. During a membership drive the same number of boys and girls joined the club. How many members does the club have if the ratio of boys to girls is now 3:4?

Answer: The club now has $3 + x$ boys and $5 + x$ girls since the same number of each sex joined.

$(3 + x)/(5 + x) = 3/4$

$4(3 + x) = 3(5 + x)$

$12 + 4x = 15 + 3x$

$4x - 3x = 15 - 12$

$x = 3$

Thus, there are now $3 + 3$ (or 6) boys and $5 + 3$ (or 8) girls.

Remember to perform a quick check to verify your answer. Interposing this step is important in decreasing careless errors.

Now, to check whether $6/8 = 3/4$:

$6 \times 4 = 8 \times 3$

$24 = 24$ (Correct answer)

More simply, does $6/8$ reduce to $3/4$? Since it does, the answer is correct.

14

Statistics

INTRODUCTION

Most math tests have some statistics questions. The concepts are relatively simple, and with a short review there is every likelihood that a student can get all of these correct. One of the main things to remember is the definition of each concept. These must be clear to answer the questions correctly.

Another thing to remember is that finding *your* answer is no guarantee that you have found the *correct* answer. The test makers are careful to choose incorrect answers that are the result of a common incorrect association or incorrect calculations. For example, if they ask for a mean and you calculate a median, you are likely to find that incorrect answer!

RELEVANT CONCEPTS FOR ALL TESTS

There are three concepts to remember that are grouped in math under the title "Measures of Central Tendency." All of these have one purpose: summarizing a set of numbers into one number that is representative of the group according to different "rules."

Mean or average = $\Sigma X/n$. That is, the sum of the numbers divided by the total number of numbers or terms.

Median is the "middle" number of a set of ordered numbers. This ordering may be from smallest to largest or vice versa; it makes no difference, the answer is the same.

Mode is the number that occurs most frequently. Note that this is not a calculation but simply an observation. A problem with the mode is that it is possibly a very extreme number and thus both less "central" and less "representative." It is also perfectly possible to have more than one mode. A bimodal distribution has two; if there are more than two modes it is termed *multimodal*.

Calculations

Mean: Given the numbers 5, 22, 17, 14, and 12, find the mean.

Answer: mean = $\Sigma X/n$, and thus $70/5 = 14$. The mean is 14.

Median: There are two different approaches depending on the number of observations or data points. If the number is odd, the process is simpler. Just choose the number in the "middle." If the total number of numbers is even, choose the two middle numbers, add them, and divide them by 2 (take the mean).
Steps for finding a median:

1. Place the numbers in order, from smallest to largest (or vice versa).
2. Is the total number of numbers odd or even?
3. If odd, select the "middle" number.
4. If even, select the two "middle" numbers, add them, and divide them by 2.

Calculation (odd): Given 8, 5, 3, 12, and 11, find the median.

Answer: Placing the numbers in order from smallest to largest, you have 3, 5, 8, 11, and 12. The "middle" number is 8, which is the median.

Calculation (even): Given 8, 5, 3, and 11, find the median.

Answer: Placing the numbers in order from smallest to largest, you have 3, 5, 8, and 11. The middle numbers are 5 and 8; their sum is 13 and the mean is $13/2 = 6.5$, which is the median.

Mode: There is no calculation for this. Simply count the number of times that numbers are repeated. In a distribution where there are no repeated observations there is no mode.

Calculation: Find the mode of 5, 8, 11, 22, and 8.

Answer: The only number repeated is 8, so that is the mode. Sometimes the calculations will not be so straightforward. Here is a good example. The concepts are exactly the same but the procedure changes slightly.

Question: Three children have a mean weight of 51 lbs. The first child weighs 60 lbs. and second child weighs 57 lbs. What does the third child weigh?

Answer: Using the formula for determining the mean ($\Sigma X/n$), find the sum of the weights.

$51 = \Sigma X/3$

$(3)51 = 3(\Sigma X/3)$

$153 = \Sigma X$

So, together, the three weigh 153 lbs. The two known weights are 60 and 57, for a total of 117 lbs. By subtraction, the third child's weight is found: 153 – 117 = 36.

A common mistake might be to take the three given numbers (51, 60, and 57) and calculate the mean. This would be found among the incorrect answers.

Another way to ask questions involving mean, median, and mode is to combine concepts. Here is an example:

Question: Five numbers have the following characteristics:

Mean = 20

Median = 20

Mode = 22

What are the numbers?

Answer: Two of the numbers must be 22 because that is the mode. There can be no other repeats. The median is 20 and with an odd number of numbers (five), that must also be included. Now we have three of the five numbers: 20, 22, and 22. The sum of these numbers is 64 (20 + 22 + 22). The five numbers must total 100 because

Mean = $\Sigma X/n$

$20 = \Sigma X/5$

$5(20) = (\Sigma X/5)5$

$100 = \Sigma X$

The last two numbers must have a sum of 36 (100 – 64 = 36) but they cannot both be 18 because the mode would change, so they must be 17 and 19. Now we have all five numbers: 17, 19, 20, 22, and 22.

However, to get this answer the test taker must understand all three concepts (mean, median, and mode) as well as be able to work with a formula in an unconventional way.

Example: During track training, Karen has the following times in seconds: 66, 57, 54, 54, 64, 59, and 59. Her three best times for the week are averaged to determine her seeding for the track meet. What is that average?

Answer: The best times (lowest numbers) are 57, 54, and 54. (Remember here the "best" times will be the *lowest* numbers. If you mistakenly choose the largest numbers you will find that answer also, but it will be wrong!)

Mean = $\Sigma X/n$

$= (57 + 54 + 54)/3$

$= 165/3$

$= 55$ seconds

Example: A salesperson drives 2,052 miles in six days, stopping at two towns each day. How many miles does this person average between stops?

Answer: The average number of miles per day is 2,052/6 or 342 miles. Since two stops are made each day, the average number of miles between stops is 342/2, or 171 miles.

Note: On most tests 342 will be listed as one of the alternatives to see if you will take the option of the first calculated answer. Be careful.

Example: What is the median of the following numbers?

14   10   20   25   14   16

Answer: The first step is to order the numbers (in this example, from smallest to largest).

10   14   14   16   20   25

Since there is an even number of numbers (6), the median will be the average of the two middle numbers. Therefore, $(14 + 16)/2 = 30/2 = 15$, which is the answer.

Example: Sarah walks to work each day following a different route because of construction. On Monday she walked 1.2 miles, on Tuesday .75 miles, on Wednesday 1.68 miles, and on Thursday .75 miles. What was the mean distance for the four days?

Answer: Since the formula for determining the mean is ÓX/n, she will have $(1.2 + .75 + 1.68 + .75)/4 = 4.38/4 = 1.095$ miles.

END OF PRAXIS I, 0014, 0511, AND 0146

Important Shortcut

To start, let's look back at the question above regarding child weights. There is a much simpler solution based on the fact that in a mean, the sum of the deviation scores is zero. These deviation scores (DS) are derived by subtracting the mean from each number. Therefore, the child weighing 60 lbs. has a DS of +9 lbs (60 – 51) and the child weighing 57 lbs. has a DS of +6 lbs. Together they have a DS of +15 lbs., so the last child must have a DS of –15 lbs., or 36 (51 – 15) lbs. This calculation can be done easily right on the test sheet, and the result can be checked quickly.

Now let's move on to some slightly harder questions in the mean/median/mode category.

Example: Sarah's average (mean) on four tests is 80. What grade must she make on the fifth test to raise her average to 84?

Answer: Sarah already has 80 × 4, or 320 points on the first four tests. With five tests, she will need 420 points (84 × 5) in order to achieve an 84 average. This is exactly 100 points more than she now has. This then is the grade needed to bring her average up to 84.

Here is a similar one where logic must be applied.

Example: Julie's average on four tests is 80. Which of the following CANNOT be the number of tests on which she had exactly 80 points?

0   1   2   3   4

Answer: You need to test each possibility logically.

Could it be 0? Yes, there are many ways four test scores could sum to 320 without one being 80.

Could it be 1? Yes, there would be many ways four scores could sum to 320 if one were 80.

Could it be 2? Yes, four numbers could sum to 320 even if two were 80.

Could it be 3? No, because if three were 80, their sum would be 240. Which only leaves 80, giving four identical numbers of 80.

Could it be 4? Yes, if all four were 80, the sum would be 320 and the mean would be 80.

Weighted Mean

This is a twist on the straightforward mean discussed above. A weighted mean has different values assigned to the component numbers. A usual example that most students understand is that of a course where quizzes, exams, and homework carry different weights.

Formula: $$\frac{w_1 x_1 + w_2 x_2 + w_3 x_3 \ldots}{w_1 + w_2 + w_3 \ldots}$$

Example: In a science class, the teacher has given the students the following schedule for determining grades for the course:

3 tests (20% each)   60%
1 paper   20%
2 quizzes (5% each)   10%
Class participation   10%

Matt has the following percentage grades:

Test scores   93, 82, 89
Paper   95
Quizzes   86, 96

What must his class participation grade need to be in order to guarantee a grade of 90% for the course?

Answer:				
3 tests	20% each	60%	weight $4 \times 3 = 12$
1 paper	20%	20%	weight 4
2 quizzes	5% each	10%	weight $2 \times 1 = 2$
Class participation	10%	10%	weight 2
		100%	20 (number of 5s)

This was set up using 5% as the weight unit because the lowest amount (lowest common multiple) was 5% (each quiz).

Test scores   93, 82, 89
Paper   95
Quizzes   86, 96
Class participation ?

Calculation:

$$90 = \frac{4(93) + 4(82) + 4(89) + 4(95) + 86 + 96 + 2X}{12 + 4 + 2 + 2}$$

$$90 = \frac{372 + 328 + 356 + 380 + 86 + 96 + 2X}{20}$$

$$90 = \frac{1{,}618 + 2X}{20}$$

$$90(20) = 1{,}618 + 2X$$

$$1{,}800 = 1{,}618 + 2X$$

$$1{,}800 - 1{,}618 = 2X$$

$$182 = 2X$$

$182/2 = X$

$91 = X$

Matt must have at least a 91 on class participation to have a final grade of 90.

Alternative Method of Weighting Using 100 Instead of 20: The advantage of 20 is that the numbers stay smaller but using 100 is more intuitive. The answers are the same.

Answer:				
3 tests	20% each	60%	weight $20 \times 3 = 60$
1 paper	20%	20%	weight 20
2 quizzes	5% each	10%	weight $5 \times 2 = 10$
Class participation	10%	10%	weight 10
		100%	100

$$\frac{93(20) + 82(20) + 89(20) + 95(20) + 86(5) + 96(5) + x(10)}{100}$$

$$\frac{1,860 + 1,640 + 1,780 + 1,900 + 430 + 480 + 10x}{100}$$

$$\frac{8,090 + 10x}{100}$$

$80.9 + .1x = 90$

$.1x = 90 - 80.9$

$.1x = 9.1$

$x = 91$

Measures of Dispersion

All three concepts here (range, variance, and standard deviation) have a single purpose, which is to express how "spread out" or dispersed a set of numbers is. Hence the name *Measures of Dispersion*.

Range: This is the distance between the most extreme observations (highest and lowest). It may be presented in two ways:

1. Name the two extremes.
2. Calculate the difference between the two extremes.

Example: My science grades are 67, 98, 83, 90, and 75. What is the range?

Answer: the range is either 98 to 67, or it is 31 (98 – 67).

Standard Deviation (from the mean): This is conceptually the average distance of a set of observations from the mean. Standard deviation (SD) has no sign but can be thought of as having ± in front because the SD extends both ways from the mean (see the normal curve below).

The manual calculation is a bit tedious, but many calculators will do it with ease.

Variance: This is the square of standard deviation. Many formulae use it as a component in other calculations, such as reliability.

Example: If the standard deviation of a distribution is 10, what is the variance?

Answer: $V = SD^2$
$V = 10^2$
$V = 100$

Example: Or more conceptually, can variance ever be less than standard deviation ($V = SD^2$)?

Answer: Yes, whenever the standard deviation is less than one. Thus if SD = .50, then V = .25 ($.50^2 = .25$).

Normal Curve

A further area where questions could fall is that of the normal curve.

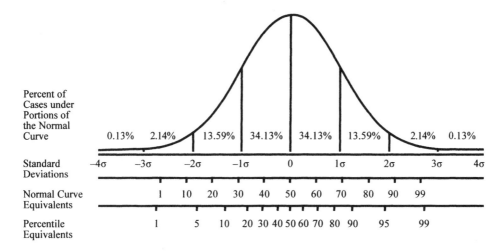

If you look carefully, you will see that it is symmetrical with observations most frequent in the center. The percents in the areas are constants and can be memorized in their whole number form: 34, 14, and 2.

Example: Given a large number of test scores distributed normally, what percent of people will have scores within one standard deviation of the mean?

Answer: 34% + 34% = 68%

Example: Given a test with a mean of 80 and a standard deviation of 5, what percent of people will have scores less than 85?

Answer: Less than 85 will be 34% + 34% + 14% + 2% = 84%

Example: Same as the last one but change to "scores greater than 85."

Answer: 14% + 2% = 16%

Example: Could be a constructed response for a test like PRAXIS 0069. Here are the results of a phone survey of 50 households.

Question: How many cars are in the households?

Responses: 7 said 0
8 said 1
21 said 2
7 said 3
7 said 4

Draw a circle graph representing the above data and calculate the angles for each segment. Show your work. Label the segments.

1. No cars: 7 responses = 360 (7/50) = 50.4 degrees
2. One car: 8 responses = 360(8/50) = 57.6 degrees
3. Two cars: 21 responses = 360(21/50) = 151.2 degrees
4. Three cars: 50.4 degrees
5. Four cars: 50.4 degrees

Results of Phone Survey of Number of Cars per Household

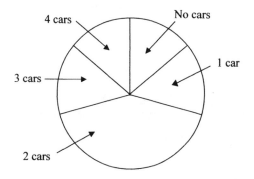

Example: From the preceding example, calculate the mean number of cars per household.

Number of cars: Having 0 0 cars
Having 1 8 cars (8 × 1)
Having 2 42 cars (21 × 2)
Having 3 21 cars (7 × 3)
Having 4 28 cars (7 × 4)
Total 99 cars

Since 50 households were surveyed, the mean is 99/50 = 1.98 cars per household.

Example: A student made the following test scores. Also shown are the means and standard deviations for each test. On which test did the student have *relatively* the lowest performance? Explain your answer.

On which test did the student have *relatively* the highest performance? Again, explain your answer.

Subject	Test Score	Mean	Standard Deviation
Science	84	90	4
Social Studies	70	78	6
English	78	84	5
Art	85	90	6

Answer: To get a handle on relatively low performance, which is different from low performance, we need to use a measure of *relative* distance from the mean. That is provided by z scores, which are calculated as follows: (score – mean)/standard deviation.

Science = (84 – 90)/4 = –6/4 = –1.5
Social Studies = (70 – 78)/6 = –8/6 = –1.33
English = (78 – 84)/5 = –8/5 = –1.6
Art = (85 – 90)/6 = –5/6 = –.83

So the subject with the largest negative z-score is English, where this student had relatively the lowest performance. Art would be the subject with relatively the highest performance (z-score = –.83).

Important Note: You can be quite sure this question will be set up so the lowest score will *not* be the answer because the test makers design questions so that using the incorrect understanding *will not* give the correct answer. The data above illustrate clearly that Social Studies has the lowest score but is not relatively the lowest.

15

Geometry, Basic Angles, and Figures

INTRODUCTION

Let's start with basic angles. Angles are measures of rotation, and rotation is measured in degrees (denoted °). A complete rotation is 360° and a half rotation is 180°. Both of these numbers are important to remember.

RELEVANT CONCEPTS FOR ALL TESTS

Point

Before a geometric shape can be formed, you must start with a point. A point has no size or dimension, yet it provides a location. An example would be a corner of an object (a box, a room, etc.), a point of origin in construction, a simple pencil point placed on paper to indicate a beginning, or a period to indicate the ending of a thought. Its notation is written with a capital letter that identifies the point.

• A

So, to construct lines, line segments, rays, and planes, there have to be at least two points, indicating a beginning and an ending. An infinite number of points join together to construct any of the above.

Line

A line has dimension but no thickness or width (i.e., it is one dimensional). It is composed of an infinite number of points that go endlessly in both directions. It is usually denoted by two points on a line (e.g., $\overleftrightarrow{AB}$ or $\overleftrightarrow{BA}$) or by a lower-case letter at one end of a line.

This line has an infinite number of points that are not visible. The only points that are showing are the ones that are needed to identify a geometrical form.

Line Segment

A line segment has dimension but no thickness because it is a *portion* or a *section* of a line. It is usually indicated by two letters. Below, BC is a line segment named by its endpoints. There are 12 line segments that can be named from the line below. A few examples are AB, AC, or AD. We could also have BA, CA, or DA in reverse notation. We can do this because end points do not indicate direction.

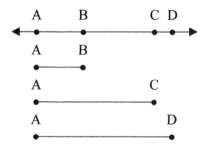
Notice that each line segment is a different length, depending on need.

Ray

A ray has dimension but no thickness or width because it is also a portion or section of a line and is usually indicated by two letters; however, it goes infinitely in *one* direction. A ray is not commutative (i.e., it cannot be named in reverse). The end point is always named first. It tells the direction of the ray.

These two rays are *not* the same. Ray $\overrightarrow{AB}$ is going to the right endlessly and ray $\overrightarrow{BA}$ is going to the left endlessly. Remember, the end point names the direction of the ray and is always named first.

Plane

A plane has a flat surface and it is two dimensional, but it has no thickness and extends infinitely in every direction, which means that it is boundless. In order to have a plane you must have at least three non-linear points.

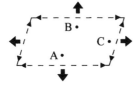 A plane's notation can be ABC or it can be named by a single letter that does *not* have a point following the letter. For example, the plane to the right could be named G.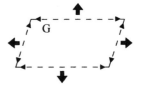

Following is a construction using points, lines, line segments, and rays on a plane.

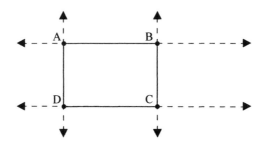

We are constructing a small building that has dimensions of 4 meters by 6 meters. We started with a point at A and formed a line to B, making a line segment AB with a length of 6. We then formed another line through points B and C, making another line segment, BC, with a length of 4. This process continued until the occupied area ABCD was identified. ABCD is a plane—a flat area that can grow as big as you want it depending on the dimensions you are seeking.

Angles

Any angle is formed from two rays extending out from a common end point, called a *vertex*.

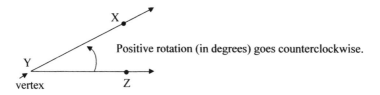

An angle can be named in three different ways. Here the angle is named either XYZ or ZYX. In either case the vertex will be in the middle to make the angle easily identifiable. It could also be identified by its vertex (in this case, Y).

There is one other way to identify an angle using numbers. This way is useful when you have more than one angle sharing a common vertex and side. These angles are called *adjacent angles*.

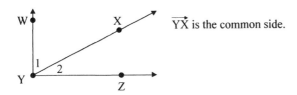

$\overrightarrow{YX}$ is the common side.

Adjacent Angles

Remember that adjacent angles are at least two angles sharing a common vertex and a common side.

Intersecting Lines

Two lines that intersect at one point (a common vertex) form four angles, called *vertical angles*. Vertical angles have four pairs of adjacent angles. They are listed below.

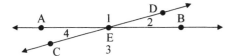

Angle AED is adjacent to angles DEB and AEC, or angle 1 is adjacent to angles 2 and 4.

Angle AEC is adjacent to angles AED and CEB, or angle 4 is adjacent to angles 1 and 3.

Vertical angles have opposite congruent angles. This means that angles 1 and 3 are equal and that angles 2 and 4 are equal.

Notice that angles 1 and 3 are obtuse and angles 4 and 2 are acute. This will always be true with vertical angles.

Perpendicular Lines

Two lines that meet to form four 90° angles may also be called *right angles*. Although perpendicular lines intersect at one point, they have different properties. They *do not* have two obtuse angles and two acute angles like vertical angles. Instead, they have four right angles that form four pairs of adjacent angles.

Note: The little square box at the point of intersection (see the illustration on the next page) indicates that each angle is a right angle. If you do not see the box, then it may be very close to a right angle but not one. That would mean the intersecting lines are not perpendicular. Instead, the line would form vertical angles. *Do not assume lines are perpendicular without evidence.*

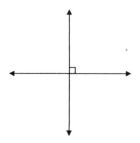

Perpendicular lines.

Complementary Angles

Complementary angles are two or more angles whose sum is 90°.

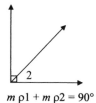 No matter how many angles are considered, if they are complementary, the sum of the measurement (m) of the angles (ρ) will equal 90°.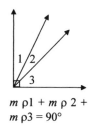

$m\ \rho1 + m\ \rho2 = 90°$

$m\ \rho1 + m\ \rho2 + m\ \rho3 = 90°$

 Angles do not have to be adjacent to be complementary. These are two separate angles but their sum equals 90°; therefore, they are complementary.

Supplementary Angles

Supplementary angles are two or more angles whose sum is 180°.

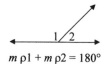 No matter how many angles are made, if they are supplementary, the sum of the angles will equal 180°.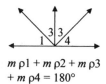

$m\ \rho1 + m\ \rho2 = 180°$

$m\ \rho1 + m\ \rho2 + m\ \rho3 + m\ \rho4 = 180°$

Angles do not have to be adjacent to be supplementary. These are three separate angles but their sum equals 180°; therefore, they are supplementary.

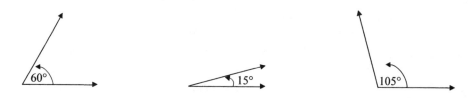

Parallel Lines

Two lines that are parallel remain equidistant at all points and never meet. Parallel lines do not form adjacent angles by themselves, but when they are combined with a third line (*m*) that intersects the parallel lines, many kinds of angle relationships are created. This third line (*m*) is called a transversal. The notation for parallel is ↔.

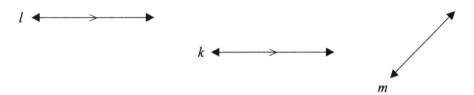

Note: Remember that a line extends infinitely, so if lines *l* and *k* kept going they would travel an equal distance from each other because the first statement says they are parallel. *Lines do not have to be directly above each other to be parallel lines; they only have to be an equal distance from each other.* If the parallel lines continued to travel toward line *m*, then they will be cut, creating eight angles.

Parallel Lines Cut by a Transversal

Same as "Parallel Lines" above, but the third line cuts across both of the parallel lines, creating eight angles. Let's put everything together using the model.

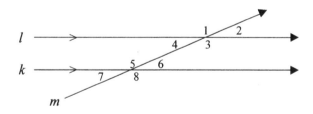

1. Look at the first parallel line (*l*). It has a straight angle equaling 180°. This means that angles 1 and 2 are supplementary.

2. Look at angles 1 through 4. If angles 1 and 2 are supplementary, then angles 4 and 3 are also supplementary. This also means that angles 2 and 3, as well as angles 1 and 4, are supplementary.

3. Notice that there are four pairs of vertical angles. What do we know about vertical angles? Non-adjacent angles are congruent. This means that angles 1 and 3 are congruent and angles 4 and 2 are congruent.

4. Since the transversal intersects both parallel lines at the same angle, then angles 5 through 8 must have the same relationship as angles 1 through 4.

This is important because, knowing this, we can know all the angles by knowing only one angle. See below to understand why.

Example: Given angle 1 with a measurement of 125°, we can find the other angles.

If angle 1 equals 125°, then angle 2 equals 55° (supplementary angles)

If angle 1 equals 125°, then angle 3 equals 125° (vertical angles)

If angle 2 equals 55°, then angle 4 equals 55° (vertical angles)

From here let's see how the second set of angles relates to the first four angles.

If angle 1 equals 125°, then angle 5 equals 125° (a transversal cutting parallel lines creates congruent relationships among angles in the *same position*).

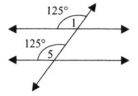 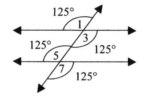 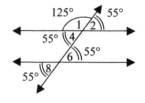

Angles 1 and 5 are called *corresponding angles*; this means that their measurements are the same.

Angles 1 and 3 are called *vertical angles*, which means that they are congruent. Angles 3 and 7 are in the same position; this means that they are *corresponding angles*, so their measurements are congruent.

Angles 1 and 2 are supplementary angles. This means that we can find angle 2 by subtraction: 180° − 125° = 55°. Since angles 2 and 4 are *vertical angles*, angle 4 is also 55°. If this is true, then angle 8 is congruent with angle 4 because they are *corresponding angles*; therefore, angle 8 is also 55.° One other angle to find is angle 6, which will also be 55° because angles 2 and 6 are corresponding angles.

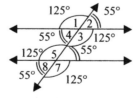

Now we have the value of all the angles by using the supplementary, vertical, and corresponding angle theorems.

Why do these relationships work?

Any time you have a set of parallel lines, they will be the same distance apart. When another line crosses them, it will duplicate angles from one parallel line to the other. These lines can now be thought of as straight angles—180°. From there we are given properties that will always be true to help solve for the first four angles, therefore duplicating the angles in the second set of angles surrounding the second parallel line.

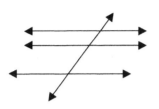

Note: This does not mean that if you have more than two lines that are parallel they have to be equal distances apart. You could have the first two lines 5 cm. apart at all points and have the third line 8 cm. from the second line and 13 cm. from the first line; however, *their angles* will always have corresponding and vertical angle relationships.

Other relationships to think about:

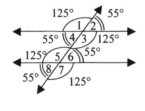

Note: Alternate *interior* angles are congruent
ρ4 and ρ6, ρ3 and ρ5
and
alternate *exterior* angles are congruent
ρ1 and ρ7, ρ2 and ρ8

Note: Same side *interior* angles are supplementary
ρ4 + ρ5 = 180°
ρ3 + ρ6 = 180°
and
same side *exterior* angles are supplementary
ρ2 + ρ7 = 180°
ρ1 + ρ8 = 180°

The sum of angles 1, 2, 3, and 4 is 360° and the sum of angles 5, 6, 7, and 8 is 360°, and the sum of angles 1, 4, 5, and 8 is 360° and the sum of angles 2, 3, 6, and 7 is 360°. Knowing these properties will help you find solutions to many geometrical problems. Not only should you know their relationships but also the proper terminology. There may be constructed response questions that will use these terms or relationships for clues to solve test questions.

What happens when a transversal cuts two or more lines that *are not* parallel?

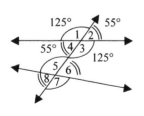

Notice that the first four angles can be still be found if one angle is given; however, we cannot determine the value of angles 5 through 8 because the angles are not in the same relationship.

By the same token, angles 5 through 8 can be determined if one angle is given; however, we cannot determine the first four angles from these.

The same terminology is still used to identify the location of the angles. For example, angles 1 and 5 are still corresponding angles, angles 4 and 5 are still same side interior angles, and each set of 4 angles will have a sum of 360°.

Here are the basic classifications of angles.

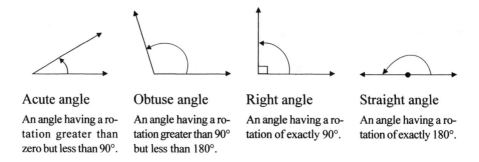

Acute angle
An angle having a rotation greater than zero but less than 90°.

Obtuse angle
An angle having a rotation greater than 90° but less than 180°.

Right angle
An angle having a rotation of exactly 90°.

Straight angle
An angle having a rotation of exactly 180°.

PRACTICE EXAMPLES

Example: In the figure below, AB and CD intersect. If angle 2 = 53°,

1. What are the number of degrees for angle 4?
2. What types of angles are 2 and 4?
3. What is maximum number of acute angles and obtuse angles that will be in vertical angles? Explain.
4. How many supplementary angles are in vertical angles?

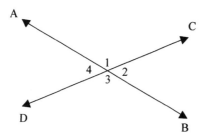

Answers:

1. Since vertical angles are congruent (in layman's terms, "opposite angles are equal"), angle 4 is 53°.
2. Angles 2 and 4 are acute angles.
3. There are two acute angles and two obtuse angles. Since each intersecting line is the same as a straight angle (180°), each pair of angles is supplementary; therefore, one angle will be greater than 90° and the other will be less than 90°, and their sum will be 180°. This means that one is an obtuse angle and the other is an acute angle. Now, since vertical angles are congruent, the maximum number of acute angles will be two and the maximum num-

ber of obtuse angles will be two. Since there are only four angles and the opposites are congruent, then an acute angle has to be opposite another acute angle and the other two opposite angles have to be obtuse.

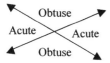

4. There are four supplementary angles: ρ1 + ρ2, ρ2 + ρ3, ρ3 + ρ4, and ρ4 + ρ1.

Example: Find the complement of 17°.

Answer: The complement of 17° is 90° − 17°, or 73°.

The complement of t° is (90 − t)°. Since the number of degrees in t is unknown, the closest you can come to defining its complement is to say 90 − t°.

Example: Find the supplement of:

1. 124°

2. (x + 9)°

Answer for #1:

The supplement of 124° is (180° − 124°) = x°

$$56° = x°$$

Answer for #2:

The supplement of (x + 9)° is 180 − (x + 9)°

$$(180 − x − 9)°$$
$$(171 − x)°$$

POLYGONS

The characteristics of polygons are:

- They are plane figures (two-dimensional shapes).
- They have at least three sides.
- They are closed figures.
- Their sides do not intersect.
- They have only line segments for sides (not arcs).

From the above characteristics, which of the following shapes are polygons?

This is not a polygon because all sides *are not* line segments.

This is the only one that is a polygon.

This is not a polygon because the sides intersect.

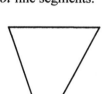

This is not a polygon because there is an opening (i.e., it is not a closed figure).

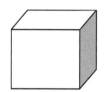

This is not a polygon because the shape is three-dimensional.

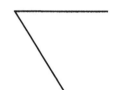

This is not a polygon because it does not have at least three sides and it is not closed.

TRIANGLES

The polygon with the fewest sides is the triangle because that is the fewest number of sides that can create a closed figure.

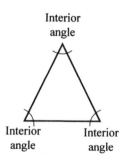

Properties of a Triangle

A triangle has three sides and the sum of its interior angles is always 180°. How did mathematicians determine the sum of the interior angles of a triangle? If you take a triangle and open it all the way, it will make a straight angle. Remember, a straight angle equals 180°.

Another way is to cut each vertex from a triangle and connect the vertices at one point and you will have a straight angle.

straight angle—180°

Now let's relate what we have already learned to some questions.

Example: If the sum of the interior angles of a triangle is 180°, what type of angles do triangles have?

Answer: Supplementary.

Three Types of Triangles

1. *Equilateral*—three equal sides and three equal angles (60°).

Acute only

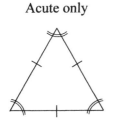

No matter how large or how small an equilateral triangle is, it will *always* have three 60° angles.

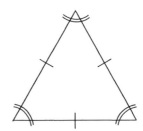

2. *Isosceles*—A triangle with two equal sides and two equal angles. An isosceles triangle has certain names to identify its parts. The two congruent sides are called the *legs* and the two congruent angles are called the *base angles*. The angle formed by the legs is called the *vertex angle* and the side opposite the vertex angle is called the *base*.

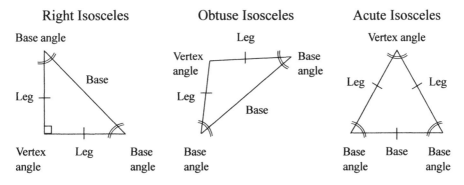

3. *Scalene*—*No* equal sides and *no* equal angles.

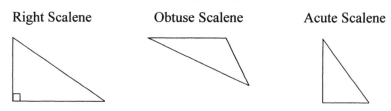

Important Facts about Triangles

1. One side of a triangle will always be smaller in measurement than the *sum* of the other two sides. For example: side = 9, side = 2, side = 5.

$$\left.\begin{array}{l} 9 + 2 > 5 \\ 9 + 5 > 2 \\ 2 + 5 \not> 9 \end{array}\right\}$$ All three sides have to follow the rule in order for it to be a triangle. If any one side is greater than any two sides, it will not be a triangle. Since the sum of side = 2 and side = 5 is not greater than side = 9, these measurements *will not* form a triangle.

A triangle *cannot* have more than one right angle in its construction; it will have no solution. See the illustration above.

A triangle *cannot* have more than one obtuse angle in its construction; it will have no solution. See the illustration above.

2. Being able to identify corresponding parts *of congruent* triangles is an important strategy. Tests will have diagrams with some information; however, if you know how to identify corresponding parts from *one triangle to another*, you could easily find the solution. The situation may involve finding the distance across a river or bridge. It's a distance that is not easily measured but can be estimated by a triangle with the same corresponding parts.

Corresponding angles are angles in the same position that have the same measurement. It is like having one triangle placed directly on top of another triangle with the triangles having the exact same angles at all vertices. Angle A corresponds to angle D, angle B corresponds to angle E, and angle C corresponds to angle F.

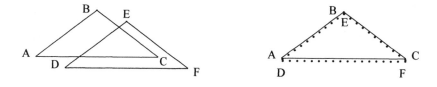

Corresponding sides are sides in the same position that have the same measurement. Using the preceding diagram, side AB corresponds to side DE, side BC corresponds to side EF, and so on.

If both triangles are *congruent*, all corresponding angles and all corresponding sides will be equal. So if triangle ABC is congruent with triangle DEF, then all the following will be true:

ρA = ρD	ρB = ρE	ρC = ρF
$\overline{AB} = \overline{DE}$	$\overline{BC} = \overline{EF}$	$\overline{AC} = \overline{DF}$

Note: The arrangement is important. Since we started with angle A, we have to start with the angle D in the other triangle because corresponding parts are named in the same order. For example:

Here triangle ABC corresponds to triangle DEF and $\overline{AC}$ corresponds to $\overline{DF}$, even though they are not oriented the same way.

3. From this concept are derived certain postulates—*Side-Side-Side* (SSS), *Side-Angle-Side* (SAS), and *Angle-Side-Angle* (ASA). These are tools that can help you find measurements that are not so easily calculated. There are many reasons why calculations are not so easily made—for example, a building may be too tall to measure its height or the distance between rivers may be nearly impossible to measure. The whole idea is to create a triangle that is *similar* or *congruent* to the one you are working with and then use the postulates to find the missing measurement.

How to Find Solutions Using Congruent Triangles

The SSS postulate says that if all three sides of one triangle are congruent with the three sides of the other triangle, then their *corresponding* angles are also congruent. This also means that the two triangles are congruent.

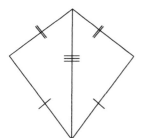

Notice that the side located in the middle is a side that both triangles share. When two triangles share a side it is referred as a *common* side.

The SAS postulate says that if two sides and the included angle of one triangle are congruent to the *corresponding* parts of another triangle, then the triangles are congruent.

Notice the vertex angles between the two sides on both triangles. This is what is meant by the term *inclusive angle*. What do we know about two lines that intersect? They form *vertical angles*. And what do we know about vertical angles? Their opposite angles are congruent. This is enough information to say that both triangles are congruent.

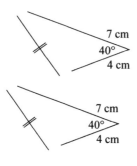

What makes this true? When two corresponding sides have the same lengths and the corresponding angles are equal, the third side is determined by the angle and the length of the sides, which means that the third side has to be the same length in order for it to form a triangle. This also means that when the third side is connected, it will form the same corresponding angles on either side of the third side as in the first triangle.

The ASA postulate says that if two angles and the included side of one triangle are congruent to *corresponding* parts of another triangle, then the triangles are congruent.

Notice this time that the indicated side is between two angles. If two corresponding angles are equal and are connected by an equal side, the third angle and both the remaining sides will be equal in order to fit properly.

How to Find Solutions Using Similar Triangles

The SSS, SAS, and ASA postulates can also be used to find *similar* triangles. This means that the side lengths will be proportional but the angles will be congruent. The "similar triangles" concept uses the concepts described in Chapter 13, "Ratios and Proportions."

The SSS postulate says that if all three sides on one triangle are proportional to the three sides of another triangle, the triangles are similar.

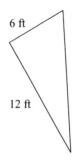

Notice that the triangles are the same shape but are not the same size. This is a clue to think, "could these triangles be similar?"

To answer the question, set up the corresponding sides into a proportion. Then ask the question, "Are these two proportions equal?" If they are, then the two triangles are proportional if the included angles are equal.

$$\frac{2}{6} = \frac{4}{12}$$

An easy way to find out is to cross multiply. You can do this because each expression is separated by an equals sign.

$$\frac{2}{6} \times \frac{4}{12}$$

$6 \times 4 = 24$

$2 \times 12 = 24$

$24 = 24$

Does this mean that the long side is 24 feet? No, but the two longer sides will be proportional. The smaller triangle is increased by a scale factor of 3 or the larger triangle is decreased by a scale factor of $1/3$.

The SAS postulate says that if the lengths of two sides of a triangle are proportional to the lengths of two sides of another triangle and the included angles of both triangles are congruent, then both triangles are similar.

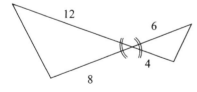

Notice that the two triangles form *vertical angles*. Now, what do we know about vertical angles? Their opposite angles are congruent. If the *corresponding sides* are proportional, this is enough information to say that both triangles are similar. Compare longest side to longest side and the other two measurements to each other.

$$\frac{6}{4} \times \frac{12}{8}$$

$6 \times 8 = 48$

$4 \times 12 = 48$

$48 = 48$

Does this mean that the long side is 48 feet? No, only that the third sides are proportional.

The smaller triangle is increased by a scale factor of 2 or the larger triangle is decreased by a scale factor of ½.

The *Angle-Angle* (AA) postulate says that if two angles of one triangle are congruent to two angles of another triangle, then the triangles are similar. (*Note*: This is different than the ASA postulate for congruent triangles. This postulate *does not* say anything about the side between the angles, as the ASA postulate does. Just because the angles are all congruent in a triangle does not *necessarily* mean the side lengths are all congruent. All equilateral triangles are similar and are congruent if the sides are equal.

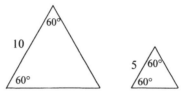

Right triangles are like isosceles triangles in that they also have special names for their parts. From the right angle extend two sides. These sides are called the *legs*, and the side opposite the right angle is called the *hypotenuse*. This will always be true no matter how the right triangle is oriented. Being able to recognize these properties will help you easily find clues in finding angles and side lengths.

One way to test whether a triangle is a right triangle is to use the Pythagorean Theorem: The sum of the squares of the lengths of the legs is equal to the square of the length of the hypotenuse.

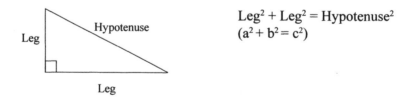

$\text{Leg}^2 + \text{Leg}^2 = \text{Hypotenuse}^2$
$(a^2 + b^2 = c^2)$

The Pythagorean Theorem *does not hold true* when a triangle is either obtuse or acute. If the square of the length of the longest side is greater than the sum of the squares of the lengths of the other two shorter sides, the triangle is obtuse. Otherwise it will be acute.

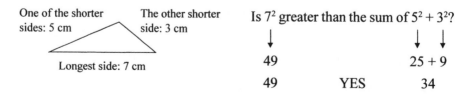

There are two triangles with special relationships that will help you solve many problems:

Equilateral triangle: Remember that this triangle has three equal angles of 60 degrees. If the triangle is bisected into two congruent triangles, it will create two right triangles. This means that each of the triangles will have angles that are 30, 60, and 90 degrees, giving special properties that will always be true.

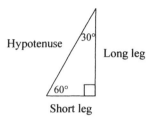

The long leg will always be the short leg times $\sqrt{3}$; the hypotenuse will always be twice the length of the short leg.

Isosceles triangle: Remember that the isosceles triangle has two sides (legs) that are congruent and two base angles that are congruent. When it is bisected, it creates two right triangles, which may be solved by conventional methods.

Another important relationship to decipher is the relationship of angles to the sides of a triangle. If you are given measurements without a clear location, you can easily place the measurements in their proper place with these simple properties of a triangle:

1. The largest angle will be opposite the longest side.
2. The smallest angle will be opposite the shortest side.

These properties will help you develop good reasoning skills.

Example: The side opposite an angle of 80° is 8 cm. The other two sides are 2 cm. and 7 cm. From this information, you can determine that 80° is the largest angle because 8 cm. is the largest side. You also can determine that all the angles will be acute because 80° is acute and it is the largest angle. You know by the side lengths that the triangle is scalene because none of the side lengths are the same and none of the angles will be equal.

Test makers are not interested in whether you know how to repeat the above information. They will give you situations intertwined with other geometrical shapes that will have little information to see if you understand how to use these properties to solve problems. If it is a constructed response question they will want not only a detailed diagram but also an explanation of how you arrived at the solution using these properties.

QUADRILATERALS

Quadrilaterals are two-dimensional figures that have four closed sides. The sum of their interior angles will always equal 360° and they will always have four angles. They are nearly the smallest polygon that can be constructed. There are several types of quadrilaterals that we learned in grade school; nevertheless, *we should not take these shapes for granted*. It is important to understand their properties because you will see them again and again. Their names are *square*, *rhombus*, *rectangle*, *parallelogram*, and *trapezium*.

The best way to show how quadrilaterals relate to each other is with a diagram:

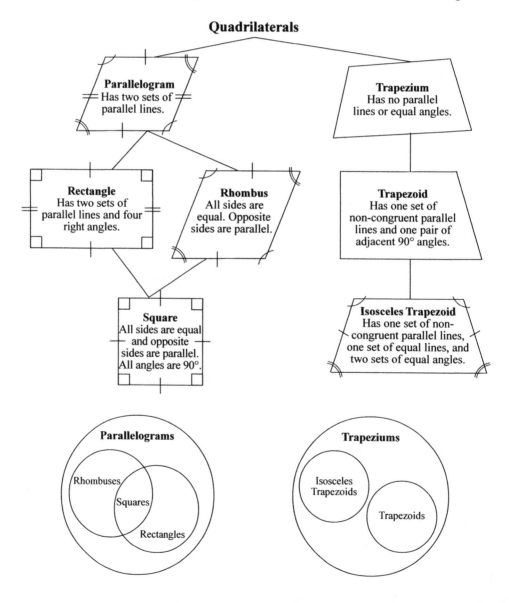

Now let's relate quadrilateral properties to what we have already learned about intersecting lines, angles, and triangles.

Parallelogram

A parallelogram has four sides and its opposite sides are parallel.

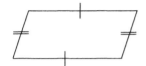
A parallelogram's opposite sides are congruent.

A parallelogram's opposite angles are congruent (two acute angles and two obtuse angles). The sum of the four interior angles is 360°.

In a parallelogram, consecutive angles are supplementary (= 180°) (e.g., 110° + 70° = 180°).

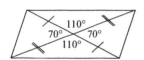
Diagonals are lines that extend to opposite vertices. Each diagonal bisects the other, which means that each segment is cut equally, but *this does not necessarily mean that both diagonals are the same length.* Diagonals also form vertical angles with a sum of 360°.

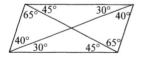
Diagonals can also be thought of as transversals going through a set of parallel lines. This means that each diagonal creates congruent alternate interior angles. Notice that the diagonals *are not* angle bisectors. This means that the angles are not equally divided.

Rhombus

A rhombus is a "special case" parallelogram that has the same properties as a parallelogram *except*:

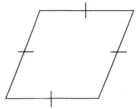 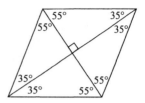

The side lengths are all congruent.   The diagonals are perpendicular, which means that the angles are bisected.

Rectangle

A rectangle is also a subset of a parallelogram in that it has two sets of congruent parallel lines. The angles are all 90° and the diagonals are congruent.

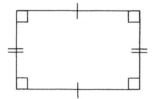 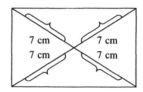

Two sets of congruent parallel lines and the angles are all 90°.   Diagonals are congruent. This means that each diagonal is bisected.

Square

A square is also a subset of a parallelogram in that it has two sets of congruent parallel lines. A square is like a rectangle in that it has four 90° angles and the diagonals are congruent, but it is like a rhombus in that all sides are equal in length and each angle is bisected.

Two sets of congruent parallel lines, the angles are all 90°, and diagonals are congruent.   All angles are bisected, which means that the diagonals bisect the vertices.

Trapezoid

Notice that the trapezoid is under the *quadrilateral* because it has four sides and the sum of its interior angles equals 360°; however, it is *not* under the branch of *parallelograms* because their properties are not at all the same. The difference is that a trapezoid has either *one* or *no* sets of parallel lines.

Isosceles Trapezoid

Trapezoid

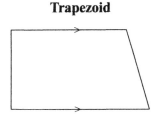

Has four sides, with one pair of sides parallel. The other side lengths are congruent. It also has two sets of congruent angles.

Has four sides, with one pair of sides parallel.

Trapezium

Has four sides, with no parallel sides and no congruent angles.

Names of Other Polygons	Number of Sides
Pentagon	5
Hexagon	6
Heptagon	7
Octagon	8
Nonagon	9
Decagon	10

HOW TO DETERMINE THE SUM OF THE INTERIOR ANGLES OF CONVEX POLYGONS

- Let n represent the number of sides.
- Let S represent the sum of degrees.
- The 2 and 180° are the constants. 180° represents the fewest number of degrees that form a polygon, so that is the *unit* measure.

△	3 sides	$S = 180(n - 2)$	$S = 180(3 - 2)$ $180(1)$ $180°$
◻	4 sides	$S = 180(n - 2)$	$S = 180(4 - 2)$ $180(2)$ $360°$
⬠	5 sides	$S = 180(n - 2)$	$S = 180(5 - 2)$ $180(3)$ $540°$
⬡	6 sides	$S = 180(n - 2)$	$S = 180(6 - 2)$ $180(4)$ $720°$
	7 sides	$S = 180(n - 2)$	$S = 180(7 - 2)$ $180(5)$ $900°$
	8 sides	$S = 180(n - 2)$	$S = 180(8 - 2)$ $180(6)$ $1080°$
	9 sides	$S = 180(n - 2)$	$S = 180(9 - 2)$ $180(7)$ $1260°$
	10 sides	$S = 180(n - 2)$	$S = 180(10 - 2)$ $180(8)$ $1440°$

Notice that the diagonals *do not cross* but intersect at one point, forming n triangles in each shape. Also, it is *not important* that each convex polygon has the same side lengths to find the *total angle* degrees, but it *is important* if you are trying to determine the *degrees for each angle*. You cannot properly find each angle degree by simply dividing the number of sides by the total degrees if each side is *not* congruent or if the polygon is concave. Remember the SSS postulate: if all three sides are congruent then all three angles are congruent as well. Look back and review the properties of equilateral, isosceles, and scalene triangles. ("Concave" means that some sides will turn inward but it is still a polygon.)

Note: The sum of all exterior angles of a *convex* polygon will always equal 360°. Using what we already know about supplementary angles, we can determine the exterior angles.

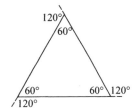 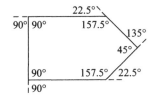 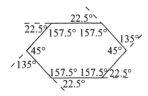

Given that each interior angle is 60°, the supplementary angle is 120°: 120 + 120 + 120 = 360.

Given the interior angles, the supplementary exterior angles can be found. Notice that their sum equals 360°: 2(22.5) + 2(90) + 135 = 360.

Given the interior angles, the supplementary exterior angles can be found. Notice that their sum equals 360°: 4(22.5) + 2(135) = 360.

PRACTICE EXAMPLES

Example: Two angles of a triangle measure 50° and 75°. How many degrees are in the third?

Answer: 50° + 75° = 125°
       180° − 125° = 55°
The third angle measures 55°.

Example: In triangle ABC, angle C is 3 times greater than angle A and angle B is 5 times greater than angle A. How many degrees are in each angle?

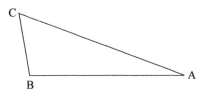

Answer: How to approach this question:
- Write down what each variable stands for.
  $C = 3A$ } angle C equals 3 times angle A
  $B = 5A$ } angle B equals 5 times angle A

- Now ask: What do I know about a triangle? It has three angles and the sum will always equal 180°.
- Knowing this information, we can set up the following equation:
  $A + 3A + 5A = 180°$
- Once the equation has been written, it will be easy to solve using a two-step process.

$A + 3A + 5A = 180°$
$9A/9 = 180/9$
$A = 180/9$
$A = 20$
So angle A is 20°
Angle C = 3(20) or 60°
Angle B = 5(20) or 100°

1. Since the variables are the same, we can add the coefficients together, making 9A.
2. Now isolate the variable to find the value of A by dividing both sides of the equation by 9.
3. Now the variable A is isolated and equals $180/9$.
4. Now divide 180 by 9 to find the value of A.
5. Substitute 20 for the variable A to find the values of the other two angles.

Does this answer check?
$20° + 60° + 100° = 180°$
$180° = 180°$
Yes, the answer checks.

HOW TO THINK ABOUT PERIMETER AND AREA FOR PLANE FIGURES

There are two types of measurements for two-dimensional figures: *perimeter* and *area*.

Perimeter measures the distance around the figure. The easiest way to find the perimeter is to add all of the sides. The problem comes when you do not know all of the side lengths. Knowing the properties of each figure is helpful in solving for the perimeter.

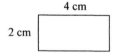

If this triangle is an equilateral triangle with one side measuring 4 cm, you could easily find the other sides by knowing that all sides are congruent; therefore, you simply add each side or multiply 3 times that value.

If only two values are given for this rectangle, you could easily find the other sides by knowing that opposite sides are congruent; therefore, you simply add the value of each side (2 + 2 + 4 + 4 = 12 cm) or you could say 2(2) + 2(4) = 12 cm. Remember, multiplication is repeated addition.

Area measures the surface inside a two-dimensional figure. This measurement is not as easy to find as perimeter. You need to know the base and height to solve for the area. The unit of measurement also has a different notation than perimeter. Perimeter is simply centimeters, inches, feet, and so on; however, area is written with the power of two following the unit—cm², ft², in². This is because we are multiplying two units of length, the height and the base, whereas with perimeter we are simply adding the side lengths. The height is sometimes tricky. It is always the altitude (the perpendicular distance from the base of the shape). The formula is base times height (notation is A = bh).

Notice that the vertical dotted line is perpendicular to the base. This is the height, and the base is either the top or the bottom of the figure. If it is given that the figure is a square, then you should know that all of the sides are congruent. To find the area of a square, multiply the base times the height—4 cm • 4 cm = 16 cm².

Notice that the vertical dotted line is perpendicular to the base. This is the height. The area formula is: A = ½ bh.
½ • 4 • 4 = 8 cm².
Notice that this area is half the area of the square in the first example. What you are really doing is getting the area of a square first, and then dividing the square into two triangles. You only want the area of one of the triangles, so multiplying by ½ gives that result.

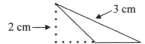

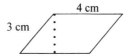

It is not so easy to get the height of this triangle. You must draw a line perpendicular from the vertex to the base *extended*. This will form two triangles, one larger and one smaller. In order to solve for the area of this triangle you will need to first find the length of the perpendicular line by using the Pythagorean Theorem, and then use that information to complete the computation using the ½ bh formula.

This parallelogram has sides that are 3 cm and 4 cm, but do not make the mistake of thinking that 3 cm is the height, because it is not. If you reconstruct the side lengths to be perpendicular without moving the distance of the two bases you will see that the original side length will extend past the bases, proving that the original side length is longer than the distance between the two bases. To find the height, you will need draw a line perpendicular from the vertex to the base and find the length using the Pythagorean Theorem before solving for the area using the formula bh.

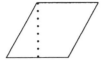 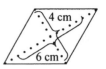

A rhombus has four equal sides but is canted over like a parallelogram. You can find the area using the same formula as for a parallelogram, base times height (bh), and like the parallelogram you may not be given the height, so you will have to find the length of the vertical leg of the right triangle first, then solve for the area of the rhombus.

Sometimes you are only given the length of the diagonals that are inside a rhombus. This is enough information to find the area when no other measurements are given. For example, let's name the diagonals d_1 and d_2. $d_1 =$ 6 cm and $d_2 = 4$ cm. Now use the triangle formula $\frac{1}{2}$ bh. You can see that we have two triangles inside the rhombus. We have the base for either one and we can derive the height of either one by taking the diagonals and dividing them by 2. Since this is true, we can simply multiply $\frac{1}{2} \cdot d_1 \cdot d_2$ to solve for the area.

$$\tfrac{1}{2} \cdot 4 \cdot 6 = 12 \text{ cm}^2$$

The area *of regular* polygons with five or more sides is not as easy to calculate. Most tests will give the area for more advanced problems; however, a polygon with six sides (hexagon) has a special relationship with the Pythagorean Theorem that allows an easy area calculation. In the example below, all that is given is the side length, which means that all six side lengths are 6 cm. How do you find the measurement of the apothem?

The *apothem*—where a line from the center point is extended perpendicularly to the base of the regular polygon.

1. Draw a diagonal from the center point to each of the vertices, as shown in the diagram below. Notice that this forms an equilateral triangle. This means that all angles are 60° and all sides are 6 cm.

2. Also notice that the *apothem* forms two right triangles inside the equilateral triangle. This gives two 30°–60°–90° triangles. Remember that the long leg is √3 times the short leg.

3. If the side lengths are 6 cm, half of that will be 3 cm, so the height of the apothem is 3√3. Now we can solve for the area by plugging in the values.

A = ½ ap (a = apothem measurement; p = perimeter calculation for the hexagon)

A = ½ • 3√3 • 36

A = 93.53 cm²

Calculating the Area of a Trapezoid

A trapezoid is unique because it has two bases. It is like a rectangle that has been squeezed at the top (base 1). When this happens, the bottom (base 2) extends the same amount that the top is squeezed; therefore the bases are added together and then divided by 2 to get the average of the two bases.

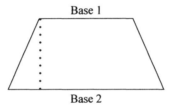

Like the other figures above, the formula is base times height but it looks a bit different because the two bases are averaged.

A = (base 1 + base 2) • height
    ───────────────────────
              2

It can also be written as A = ½ • (base 1 + base 2) • height

If the height is not given, you will have to use the Pythagorean Theorem to solve for the height before you can solve for the area. Remember that the side lengths *are not* the height.

Approximating the Perimeter and Area of a Circle

Circles are not polygons but are in the family of plane figures because they are two-dimensional. Here I will show how to approximate the perimeter and area of a circle using its components.

The center point of a circle is equidistant from all points on the circle's perimeter, or *circumference* (see the circle below, left). The arrow represents half the distance from the center point to a point on a circle. This arrow is called the *radius*.

If I take the radius and extend it through the center point from one end of the circle to the other (see the circle below, right), the line will then be called the *diameter*.

The radius and the diameter have a special relationship with the circumference. No matter how large or small the circumference is, the ratio of the diameter to the circle will always be a constant of 3.14 ...... This value has a Greek name called pi (the notation is π). This phenomenon will be important in approximating the perimeter and area of a circle.

Perimeter: For a circle, the perimeter (or circumference) equals the linear distance around the figure. The formula is

$$\pi d \quad \text{or} \quad 2\pi r$$

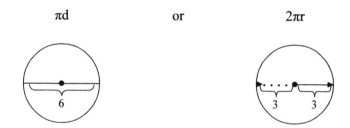

If the diameter is 6 units, I could approximate the circumference by multiplying 6 times π (3.14): 6 times 3.14 ≈ 18.85.

If the diameter is 6 units, the radius will be 3 because it is half the diameter. I could approximate the circumference by multiplying 2 times the radius times π (3.14): 3 times 2 times 3.14 ≈ 18.85.

Note: We say "approximate" when using pi because it is an irrational number that never truncates (ends) or repeats. Whether 3.1416 or $^{22}/_{7}$ is used for a calculation, the result is only a close approximation of the actual value.

Area: Area is based on the same concepts as circumference except that the result will be in square units: πr^2 or $\pi(d/2)^2$. You simply plug the radius into the formula and solve.

3.14(3)²   *Remember the order of operations*: Exponents first, then
3.14(9)   multiply.
28.26

The approximate area is 28.26 square units.
Deriving the Formula for the Area of a Circle:

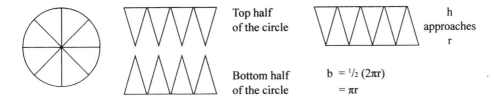

If we cut the circle in half and open up this circle, then take the top half and fit it to the bottom half, we would have a parallelogram. The arcs would form the base and the radius would be the height.

$A = b \cdot h$

$A = \frac{1}{2}(2\pi r) \cdot r$

$A = \frac{2}{2}\pi r^2$

$A = 1(\pi r^2) = \pi r^2$

Note: This approach assumes an almost infinite number of slices so that the arc approaches a straight line and the side length (radius) approaches the height.

Volume of Solid Figures

Volume calculates how much space is occupied inside a three-dimensional figure. The figures have special names: cubes, cones, prisms, cylinders, and pyramids.

Cube: A cube has three equal dimensions (length = width = height) and all 90° angles.

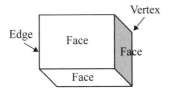

All faces, edges, and vertices are congruent. A cube has 12 edges, 6 faces, and 8 vertices.

Volume = s^3 (s • s • s)

The answer is *always* in cubic units (cubic feet, ft³, etc.), not square units (square feet) or linear units (feet).

Rectangular Prism: A rectangular prism has three dimensions with all 90° angles and the answer is *always* in cubic units.

A rectangular solid is also called a *prism*. It has two bases. They are the smaller faces that are parallel to each other and have four lateral sides.

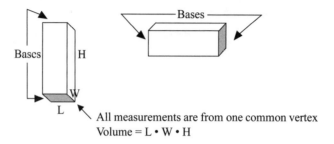

All measurements are from one common vertex
Volume = L • W • H

Example: What is the volume in feet and yards of a room with the following dimensions?

Height = 10'

Length = 15'

Width = 12'

Answer: Volume = 12 • 10 • 15

$V = 1{,}800 \text{ ft}^3$

And since 1 cubic yard = 27 cubic feet (3 • 3 • 3), the room contains $^{1800}/_{27}$ = $66^2/_3$ cubic yards.

Check: 10' ≈ 3⁺ yards

15' = 5 yards

12' = 4 yards

3⁺ • 4 • 5 = 60⁺ cubic yards; this agrees with our answer.

Example: How many cubic yards of cement will be needed to pour a basement 30 feet by 60 feet, 1 foot thick (12")?

Answer: 12" = $^1/_3$ yard

30' = 10 yards   12 inches = 1 foot

60' = 20 yards   3 feet = 1 yard

Volume = 10 • 20 • $^1/_3$ = $^{200}/_3$ = $66^2/_3$ cubic yards

Triangular Prism: A triangular prism has two triangular bases and three lateral sides that are rectangular in shape.

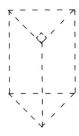 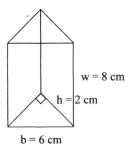

w = 8 cm
h = 2 cm
b = 6 cm

To find the volume of this prism: Find the area of the triangular base and then multiply it by the height.

A = ½(6 • 2)

A = 6 cm²

Now multiply 6 cm² • 8 cm = 48 cm³

This prism is really half of a rectangular prism (see the dashed section that is separated from the triangular prism; if the two pieces were put together they would form a rectangular prism). I could have multiplied (b • w • h)/2:

$$\frac{6 \cdot 2 \cdot 8}{2} = \frac{96}{2} = 48 cm^3$$

Hexagonal Prism: A hexagonal prism has two bases that have six sides and six lateral sides that are rectangular in shape.

2 yds

6 yds

1. Calculate the area of one of the hexagonal bases using the apothem formula (½ap).
2. Multiply the area of the base by the height of the prism to find its volume.

Because the prism has six sides, the apothem formula works beautifully to find the area of the base because of the 30–60–90 Pythagorean Theorem relationship.

Pyramid: A regular pyramid has one base that is perpendicular to one vertex. It takes three pyramids to fill in a prism. The volume formula for a pyramid is ⅓lwh or ⅓bh.

A base with four sides could form either a rectangular pyramid or a square pyramid (see the illustrations at the top of the next page).

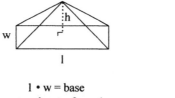

$l \cdot w$ = base
rectangle area formula
or $^1/_3(l \cdot w)h$

s^2 = base
square area formula
or $^1/_3(l \cdot w)h$

A triangular pyramid would have a base and three sides:

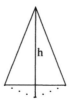

$^1/_2 bh$ = base
Then use the triangle volume formula
or $^1/_3(^1/_2 bh)h$

Cylinder or Right Cylinder: In a cylinder, sides and ends are perpendicular. They have two bases that are parallel to each other.

To find the volume of a cylinder you simply calculate the area of the circular base, then multiply it by its perpendicular height. The formula is $\pi r^2 h$.

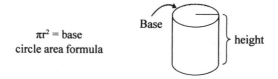

πr^2 = base
circle area formula

Cone: A cone is related to both a cylinder and a pyramid. It has a circular base and one vertex. Cones having the same radius and height would fill a cylinder three times. The volume formula for a cone is $^1/_3 \pi r^2 h$.

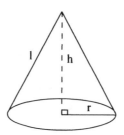

Notice that there is a lower case L down the side of the cone; this is called the *slant height*. Do not get the height or altitude and the slant height confused. When calculating the *volume* of a cone, use the height that is perpendicular to the base. If you are calculating the *surface area* of the cone section without the circular base, use the slant height formula: a = πrs where s = √r²+h². Remember that when calculating the volume, the unit of measure is always cubic units (units³).

Example: What is the volume of a cylinder with a radius of 14 feet and a height of 6 feet?

Answer: Volume = πr²h

$\quad$ V = ²²/₇ • 14 • 14 • 6

$\quad$ V = 22 • 2 • 14 • 6

$\quad$ V = 3,696 cubic feet (ft³)

Example: What are the area and perimeter of this triangle?

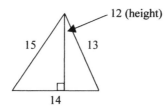

Answers: Area = bh/2

$\quad$ A = (14 • 12)/2

$\quad$ A = 7 • 12

$\quad$ A = 84 square units (not linear or cubic)

$\quad$ Perimeter = s₁ + s₂ + s₃

$\quad$ P = 15 + 13 + 14

$\quad$ P = 42 linear units (not square or cubed)

Example: What are the area and perimeter of this polygon?

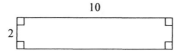

Answers: Area = lw

$\quad$ A = 10 • 2

$\quad$ A = 20 square units

Perimeter = 2l + 2w
P = 2(10) + 2(2)
P = 20 + 4
P = 24 linear units

Example: What are the area and perimeter of the polygon below?

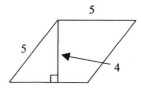

Answers: Area = bh
A = 5 • 4
A = 20 square units
Perimeter = 4s
P = 4(5)
P = 20 linear units

END OF PRAXIS I, 0014, 0511, AND 0146

The sum of *the interior* angles of a polygon can be determined by the following formula:

(n − 2)180 = sum of interior angles (where n is the number of sides)
Triangle: (3 − 2)180 = 180
Square: (4 − 2)180 = 360
Hexagon: (6 − 2)180 = 720

Example: In the figure below, the area of square ABCD is 100, and the area of isosceles triangle DEC is 10. Find the straight line distance from A to E.

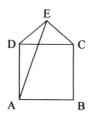

Answer: If the area of the square is 100, then each side is 10. Likewise for the triangle.

$10 = (10(h))/2$ (Using area = $\frac{1}{2}(bh)$)

$10 = 5h$

$^{10}/_5 = h$

$2 = h$

Now we can consider the triangle AEY where Y bisects line segment AB.

The hypotenuse AE can be calculated with the Pythagorean Theorem (see Chapter 18) as follows:

$a^2 + b^2 = c^2$

$5^2 + 12^2 = c^2$

$25 + 144 = c^2$

$169 = c^2$

$\sqrt{169} = c^2$

$13 = c = \overline{AE}$, which answers the question.

Example: In the figure below, BC = BE. If R represents the perimeter of rectangle ABCD and T represents the perimeter of triangle CBE, what is the value of R − T?

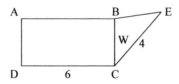

Answer: $R = 2(6) + 2(w)$

$R = 12 + 2w$

$T = 4 + w + w$

$T = 4 + 2w$

And the difference: R − T = 12 + 2w − (4 + 2w)
= 12 + 2w − 4 − 2w
= 8   Which is the answer.

Example: What is the volume of this figure?

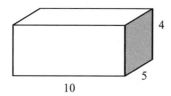

Answer: Volume = lwh

  V = 10 • 5 • 4

  V = 200 cubic units (not square, not linear)

Example: What is the volume of this figure?

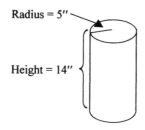

Answer: Volume = $\pi r^2 h$

  V = $(^{22}/_7)(5)(5)(14)$ or   V = $\pi(5^2)(14)$
  V = 22(25)(2)   V = $\pi(25)(14)$
  V = 1,100 cubic inches   V = 350π

Notice how using the fractional value of pi makes this calculation much easier ($^{22}/_7 \approx 3.14$).

Example: What is the surface area of the figure above? (This would be a good constructed response question for the PRAXIS Middle School Math Test.)

Answer: You need to imagine the surface area like this:

Geometry, Basic Angles, and Figures 147

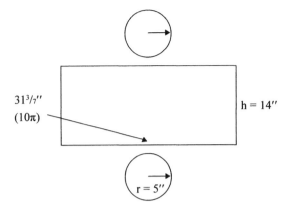

First you need to think of the surface of the side of the cylinder. The height is given, 14 inches. The width is equal to the *circumference of the circular ends*.

C = 2πr

C = 2π5

C ≈ 31³/₇ inches or C = ²²⁰/₇ inches

So the area of the *side* is:

A = lw

A ≈ 14 • ²²⁰/₇

A ≈ 2 • 220

A ≈ 440 square inches (or 140π)

And the area of each *end* is:

A = πr²

A ≈ (²²/₇)(5²)

A ≈ ²²/₇(25) ≈ ⁵⁵⁰/₇ ≈ 78⁴/₇ square inches (also 25π)

The final answer is that the surface area of the figure is:

440 + 2(78⁴/₇) ≈ 440 + 157¹/₇ ≈ 597¹/₇ square inches (597.14 in²)

The same calculation expressed in terms of pi is:

140π + 2(25π) = 140π + 50π = 190π

The Golden Rectangle

From earliest geometry in Greece, the golden rectangle has been thought of as a perfect form of a rectangle. The sides are in the approximate ratio of 1:1.618. A more exact rendition is 2:(1 + √5).

Construction of a golden rectangle is relatively easy and can be performed with a straightedge and compass. First, draw a square with sides of 2. Bisect the base side and place the point of the compass at the midpoint of that side. Now draw an arc from the upper left corner of the square to the extension of the base side. Finally, finish the rectangle. Its dimensions will be 2 by $1 + \sqrt{5}$ units.

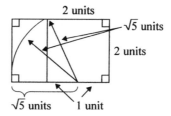

Many architects, artists, and designers consciously utilize the golden rectangle in many applications, such as room dimensions.

The golden triangle is also thought of as a perfect form. From the above rectangle you can see it within the golden rectangle, where it has the dimensions of 1, 2, and $\sqrt{5}$.

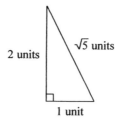

One amazing characteristic of this triangle is that four and five identical ones can be arranged to make a larger golden triangle.

Example: A room is 12′ by 19.5′. Is it a golden rectangle?

Answer: Yes, because 19.5/12 = 1.625, which is approximately $(1+\sqrt{5})/2 \approx$ 1.618 (it is only about .07 off).

16

Geometry: Perimeter

INTRODUCTION

Even though perimeter is treated in other chapters, this topic is important enough to have an entire chapter devoted to it.

Perimeter is the *linear* distance around a figure. Because it is linear, the units will *not* be either squared or cubed. The answer will be simply the distance in feet, inches, miles, meters, kilometers, and so on. Often a question will present choices that have the correct numerical value tied with the incorrect unit; don't be fooled. For example, you are given a word problem about fencing a rectangular field 50 meters by 100 meters. The sum of the lengths of the four sides is 300 (100 + 100 + 50 + 50). One of the choices is 300 meters and another is 300 square meters (or m^2). The answer must be in terms of a linear unit, so the choice in square meters can be eliminated even though the 300 represents the correct calculation. Never lose points because of this simple error.

RELEVANT CONCEPTS FOR ALL TESTS

It is simplest to imagine yourself "walking" around a figure. There are numerous perimeter formulae but each is simply a codification of how you would approach calculating that "walking around" linear distance.

Circle: = Perimeter here is termed circumference.
   Formula: $C = \pi d = 2\pi r$

Square or Rhombus: Since all four sides are equal, multiply one side length by 4.
   Formula: $P = 4s$

Rectangle or Parallelogram: Since each figure has two pairs of equal sides, perimeter is equal to 2 times one side plus 2 times the other side.

Formula: $P = 2l + 2w$

Again, just think of "walking around" the figure instead of trying to calculate a complicated formula for each.

Triangle: Since triangles can vary so much, just think that the perimeter is the sum of the three sides, even though it is entirely possible for two (isosceles triangle) or three (equilateral triangle) of the sides to be equal.

Formula: $P = s_1 + s_2 + s_3$

Word problems can include perimeter calculations when they present situations like:

Length of fence around a field

Length of baseboard for a room

Measurement of a frame

Measurement of a circular lid on a jar

Example: What length of fence is needed for a rectangular field 500 feet by 800 feet?

Length of fence (perimeter) = $2(500) + 2(800)$
$$P = 1,000 + 1,600$$
$$P = 2,600 \text{ feet (not square feet or ft}^2)$$

END OF PRAXIS I, 0014, 0511, AND 0146

Problems here are based on the same principles but are somewhat more complex.

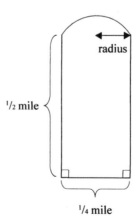

Example: A rancher is fencing a field. The straight dimensions are ½ mile and ¼ mile respectively. At one end there is a curved fence that follows a road.

Geometry: Perimeter

How much fence is needed? At a cost of $1.25/foot, how much will it cost?

Answer: Step 1: Calculate the lengths of the three straight sections.

$L = 1/2 + 1/2 + 1/4$

$L = 1\,1/4$ miles (or) $5{,}280 \cdot 1\,1/4 = 6{,}600$ feet

Step 2: Calculate the length of the curved section.

$C = 2\pi r$

$C = 2 \cdot 22/7 \cdot 1/8$

$C = 22/28 \rightarrow 11/14$ mile $\rightarrow$

$C = 11/28 \cdot 5{,}280 = 2{,}074.29$ feet

If it was a complete circle but we only need a half:

$C = (11/14)(1/2) = 11/28$ mile

Step 3: Total the first two steps.

$T = 1\,1/4 + 11/28 = 1\,7/28 + 11/28 = 1\,18/28 = 1\,9/14$ miles of fence

Step 4: Figure the cost.

$(6{,}600 + 2{,}074.29)(\$1.25) = (8{,}674.29)(\$1.25) = \$10{,}842.86$

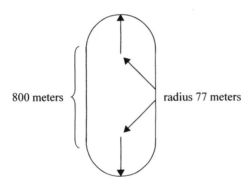

800 meters   radius 77 meters

Example: The management of an oval horse racetrack is replacing the inside rail. How much rail is needed?

Answer: Step 1: Calculate the curved ends. Their length will be equal to a circle with a radius of 77 meters.

$C = 2\pi r$

$C = 2 \cdot 22/7 \cdot 77$

$C = 2 \cdot 22 \cdot 11 = 484$ meters (or) 154π meters

Step 2: Calculate the two straight sections.

$L = 2 \cdot 800 = 1{,}600$ meters

Step 3: Find the total length of rail.

$T = 1{,}600 + 484 = 2{,}084$ meters

17

Geometry: Area

INTRODUCTION

When questions are concerned with the amount of *surface* in two dimensions, they are area questions. The answers are *always* in square units, that is, square feet (ft^2), square meters (m^2), square miles ($miles^2$), and so on; they *are never* in linear units, such as feet, meters, or miles, or in cubic units, such as cubic feet (ft^3), cubic meters (m^3), and so on. So if an option has the correct number but the incorrect unit, it is incorrect.

The questions do not have to be couched only in terms of figures like triangles or rectangles, but may be in such forms as fields, rooms, or even paint coverage.

RELEVANT CONCEPTS FOR ALL TESTS

The least demanding area problems are like this:

Example: What is the area of the following figure?

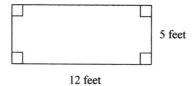

12 feet

Answer: A = lw
A = 5 • 12
A = 60 ft²

Example: Given a square with a diagonal of 6 yards, what is its area?

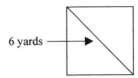

6 yards

Answer: The formula is $A = \frac{1}{2}d^2$, where d = the diagonal, here 6 yards:
$A = 6^2/2 = 36/2 = 18$ square yards (yds^2)

Be careful of questions that ask for a *conversion* of units, especially when using our customary system (inches, feet, yards, etc.).

Please be sure to remember these facts:

144 in^2 = 1 ft^2 (not 12 but 12 • 12)
9 ft^2 = 1 yd^2 (not 3 but 3 • 3)

Example: What is the area of the triangle below?

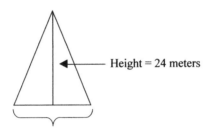

Height = 24 meters

Base = 18 meters

Answer: $A = \frac{1}{2}bh$
$A = \frac{1}{2} \cdot 18 \cdot 24$
$A = 18 \cdot 12$
$A = 216$ m^2 (square meters)

Example: Ralph is carpeting his living room with wall-to-wall carpet. How many square yards will he need?

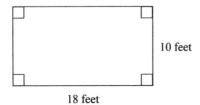

10 feet

18 feet

Answer: A = lw

A = 10 • 18

A = 180 square feet → But there are 9 square feet in one square yard.

A = $^{180}/_9$ = 20 square yards

Example: What is the area of triangle BDE? (ABCD is a rectangle)

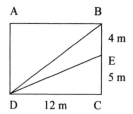

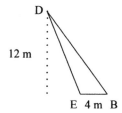

Answer: The area of a triangle is $^1/_2$ bh.

A = $^1/_2$ • 4 • 12

A = 2 • 12 = 24 m²

(Note: Think of the triangle turned 90°. Then you can see the height is 12 m.)

Example: What is the area of triangle DFH?

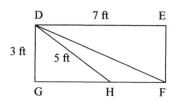

Answer: Step 1: Calculate the length of $\overline{GH}$ by using the Pythagorean Theorem ($a^2 + b^2 = c^2$).*

3² + b² = 5²

9 + b² = 25

b² = 25 – 9

b² = 16

b = 4 → $\overline{GH}$ is 4 ft.

(You could avoid this calculation by observing that the first two sides of a perfect 3-4-5 triangle would give 4 as the answer.)

Step 2: If $\overline{GH}$ is 4, then $\overline{FH}$ is 3.

Since DE is 7 ft, subtract 4 from 7 to find the length of FH:
7 – 4 = 3ft.

*Note: See Chapter 18 for more information on the Pythagorean Theorem.

Step 3: Now calculate the area.

$A = \frac{1}{2}bh$

$A = \frac{1}{2} \cdot 3 \cdot 3$

$A = \frac{9}{2} = 4\frac{1}{2}$ square feet (ft²)

Example: Vivian is making a mosaic. Each tiny piece in the artwork is $1\frac{3}{8}$ inches by $1\frac{1}{8}$ inches. What is the area of each piece?

Answer: $A = bh$

$A = 1\frac{3}{8} \cdot 1\frac{1}{8}$

$A = \frac{11}{8} \cdot \frac{9}{8} = \frac{99}{64} = 1\frac{35}{64}$ square inches (in²)

Example: Julia's lawn is 37 yards by 20 yards. Yesterday she mowed $\frac{2}{3}$ of the lawn. How many square yards are left to be mowed today?

Answer: $A = lw$

$A = 37 \cdot 20 = 740$ yds² (in total yards)

$A = \frac{2}{3} \cdot 740 = \frac{1480}{3} = 493\frac{1}{3}$ yds² (mowed yesterday)

$740 - 493\frac{1}{3} = 246\frac{2}{3}$ yds² (left to mow)

END OF PRAXIS I, 0014, 0511, AND 0146

Example: A circle is inscribed in a square that is 10 cm on each side. What is the total area of the four corner spaces outside the circle?

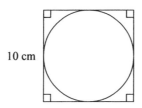

Answer: First define the desired parts in words. These pieces are the difference between the area of the square and the area of the circle.

Now that can be changed into a mathematical expression:

$A = s^2 - \pi r^2$	s^2 = area of a square
$A = 10^2 - \frac{22}{7} \cdot 5^2$	πr^2 = area of a circle
$A = 100 - \frac{22}{7} \cdot 25$	$\pi \approx \frac{22}{7}$
$A = 100 - \frac{550}{7}$	
$A = \frac{150}{7} = 21\frac{3}{7}$ cm²	

(might also be listed as $100 - 25\pi$ cm²)

Example: What is the area of ABC?

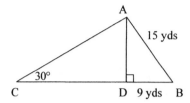

Answer: Step 1: Look at triangle ABD. The other leg is 12 yards using the Pythagorean Theorem and the perfect 3-4-5 triangle (each side multiplied by 3).

Step 2: Look at triangle ADC. Since there is one 30° angle and the short leg is 12 yards, the hypotenuse is 24 yards (30-60 rule).

Step 3: Still looking at triangle ADC, knowing two sides will give the third.

$a^2 + b^2 = c^2$

$12^2 + b^2 = 24^2$

$144 + b^2 = 576$

$b^2 = 576 - 144$

$b^2 = 432$

$b = \sqrt{432}$

$b = \sqrt{144 \times 3}$

$b = 12\sqrt{3}$

Step 4: Now we are ready to find the area. The base is $9 + 12\sqrt{3}$ and the height is 12.

$A = \frac{1}{2}(12)(9 + 12\sqrt{3})$

$A = 6(9 + 12\sqrt{3})$

$A = 54 + 72\sqrt{3}$ square yds (yds²)

Example: In the figure below, the half circle has a radius of 4 inches and the height of the triangle is 5 inches. What is the area of the entire figure?

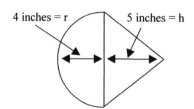

Answer: We arrive at the answer in three steps.

Step 1: Calculate the area of the half circle.

$A = \pi r^2$

$A = \pi(4^2)$

$A = 16\pi$ square inches (but the half circle would be 8π)

Step 2: Calculate the area of the triangle.

$A = bh/2$

$A = 8 \cdot 5/2$

$A = 40/2 = 20$ square inches

Step 3: Add the areas of the two figures.

$A = 8\pi + 12$ (or) $4(2\pi + 3)$ square inches

Example: Students have been assigned a class "memory" project with pictures of themselves and their families. They may use four 2-foot-by-3-foot posters. How many 4-inch-by-6-inch pictures may be mounted?

Answer: Each poster is 24 inches by 36 inches.

    Six pictures may be placed along each dimension.

    ($24/4 = 6$ and $36/6 = 6$)

    Each poster may hold $6 \cdot 6$ or 36 pictures.

    Four posters may hold $36 \cdot 4$ or 144 pictures.

Example: John is purchasing a triangular sail for his boat. The measurements are: 20-foot base with a height of 30 feet. If material is $14.99 a square yard, how much will the sail cost?

Answer: This is a three-step problem.

Step 1: Calculate the area.

$A = bh/2$

$A = 20 \cdot {}^{30}/_2$

$A = {}^{600}/_2 = 300$ square feet

Step 2: Calculate the number of square yards. Remember, there are 9 square feet in a square yard, not 3!

Yds $= {}^{300}/_9 = 33{}^1/_3$ square yards

Step 3: Now calculate the cost.

$C = 14.99(33{}^1/_3)$

$C = \$499.67$

18

The Pythagorean Theorem

INTRODUCTION

Many students know the Pythagorean Theorem ($a^2 + b^2 = c^2$) but have absolutely no idea how or when to use it. They have the *what* without any inkling of the *how* or the *when* of application. In fact, when they tell us what it is, they seem satisfied that they have actually mastered the concept. Far from true.

RELEVANT CONCEPTS FOR ALL TESTS

First of all, this theorem only relates to right triangles, that is, triangles with one 90° angle. The letters "a" and "b" refer to the sides that comprise the right angle. These are also called "legs." It doesn't really matter which one is "a" or "b." What *does* matter is that "c" always denotes the length of the longest side—also known as the hypotenuse—which is opposite the right angle.

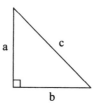

The Pythagorean Theorem: In a right triangle the square of the length of the hypotenuse equals the sum of the squares of the lengths of the legs.

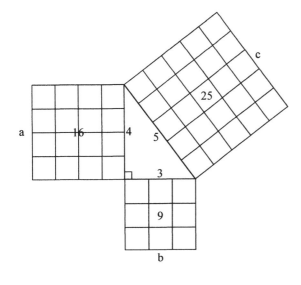

To find the length of the hypotenuse:

$a^2 + b^2 = c^2$

$4^2 + 3^2 = c^2$

$16 + 9 = c^2$

$25 = c^2$

$\sqrt{25} = \sqrt{c^2}$

$5 = c$

To find the length of side a:

$a^2 + b^2 = c^2$

$a^2 + 3^2 = 25$

$a^2 + 9 = 25$

$\underline{\ -9\ -9}$

$a^2 = 16$

$\sqrt{a^2} = \sqrt{16}$

$a = 4$

Notice that the side lengths of this right triangle are 3, 4, and 5, and that each number forms a perfect square for that side. This is what the Pythagorean Theorem is saying—that if you square each leg (a, b) it will equal the sum of the squares of the hypotenuse (c). This is a very useful tool in solving for any side length of a right triangle. All you have to do is remember the theorem $a^2 + b^2 = c^2$. Make sure the "c" variable always refers to the hypotenuse.

Another use of the Pythagorean Theorem is to see if a right triangle can be formed with three given line segments. For example, let's say the line segment measurements are 7, 10, and 9.

1. I know the longest side has to be a possible hypotenuse and the other two sides can be either leg.
2. Now plug in the values into the formula:

$a^2 + b^2 = c^2$

$7^2 + 9^2 = 10^2$

$49 + 81 \neq 100$

a^2 plus b^2 does not equal c^2, so the line segments cannot form a right triangle.

Using Pythagorean Triples will always give a side length without calculating. For example, 3, 4, and 5, 5, 12, and 13, and 7, 24, and 25 (and their multiples) will always form right triangles.

3, 4, and 5 multiples			5, 12, and 13 multiples			7, 24, and 25 multiples		
a	b	c	a	b	c	a	b	c
3	4	5	5	12	13	7	24	25
6	8	10	10	24	26	14	48	50
9	12	15	15	36	39	21	72	75
12	16	20	20	48	52	28	96	100

You can create your own Pythagorean Triples using the table below:

n	$2n = a$	$n^2 - 1 = b$	$n^2 + 1 = c$	n	$2n = a$	$n^2 - 1 = b$	$n^2 + 1 = c$
3	6	8	10	6	12	35	37
4	8	15	17	12	24	143	145
5	10	24	26	13	26	168	170

To confirm the accuracy of these tables, let's do a couple of calculations. From the left side of the table we can substitute 8, 15, and 17 for a, b, and c in the Pythagorean Theorem.

Does $8^2 + 15^2 = 17^2$?
$64 + 225 = 289$
$289 = 289$ Yes.

Let's do a row from the right side.

Does $26^2 + 168^2 = 170^2$?
$676 + 28{,}224 = 28{,}900$
$28{,}900 = 28{,}900$ Yes.

Although this is good information, the most commonly used Pythagorean Triples are 3, 4, and 5 and 5, 12, and 13.

There are two right triangles that have special characteristics that can be used to solve for side lengths without having to use the Pythagorean Theorem.

The 30°–60°–90° triangles: The short leg will always be either ½ the length of the hypotenuse or the hypotenuse will always be twice the length of the short leg and the long leg will always be $\sqrt{3}$ times the short leg (*l*).

The 45°–45°–90° triangles: Since both legs are the same length, the hypotenuse will always be $\sqrt{2}$ times the length of either leg (*l*).

The 30°–60°–90° triangles   The 45°–45°–90° triangles

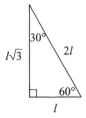 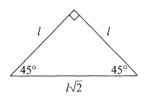

Now let's try a simple calculation.

> If the legs of a right triangle are 6 meters and 8 meters, respectively, how long is the third side (the hypotenuse)?

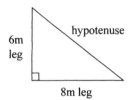

$a^2 + b^2 = c^2$

$6^2 + 8^2 = c^2$

$36 + 64 = c^2$

$100 = c^2$

$\sqrt{100} = \sqrt{c^2}$

$10 = c$

Therefore, the hypotenuse is 10 meters long.

However, the problem may not give the lengths of both legs. Let's do a problem like that.

> If the hypotenuse of a right triangle is 15 inches and one leg is 12 inches, how long is the third side (leg)?

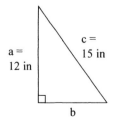

$a^2 + b^2 = c^2$
$12^2 + b^2 = 15^2$
$144 + b^2 = 225$
$b^2 = 225 - 144$
$b^2 = 81$
$b = \sqrt{81}$
$b = 9$

The other side (leg) is 9 inches long.

Another variation in the application of the Pythagorean Theorem is a two-step problem where you use the theorem to obtain data to solve the main question.

Example: An isosceles triangle has two equal sides, each 5 feet long. The base is 8 feet long. What is the area?

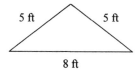

Answer: Step 1: You need to find the height of the triangle in order to calculate its area (A = bh/2). To do this you need to bisect the triangle into two equal right triangles, like this:

Step 2: Now apply the Pythagorean Theorem.
$a^2 + b^2 = c^2$
$4^2 + b^2 = 5^2$
$16 + b^2 = 25$
$b^2 = 25 - 16$
$b^2 = 9$
$b = \sqrt{9}$
$b = 3$

The height is 3 feet.

Step 3: Now find the area.

   A = bh/2

   A = 8 • 3/2

   A = 24/2

   A = 12

The area is 12 square feet.

Special note: On areas, if the answer is not in square units (square yards, square miles, square inches, etc.) it is *not* correct. In the last example, 12 feet would be incorrect.

END OF PRAXIS I, 0014, 0511, AND 0146

At this level the questions become a bit more complex.

Example: A 25-foot-tall ladder is placed against a vertical wall of a building, with the bottom of the ladder standing on concrete 7 feet from the building. If the top of the ladder slips down 4 feet, how far will the bottom of the ladder slide out?

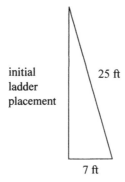

initial ladder placement   25 ft

7 ft

Step 1: Calculate the third side.

   $a^2 + b^2 = c^2$

   $7^2 + b^2 = 25^2$

   $49 + b^2 = 625$

   $b^2 = 625 - 49$

   $b^2 = 576$

   $b = \sqrt{576}$

   $b = 24$ feet

Step 2: Let the ladder slide out.

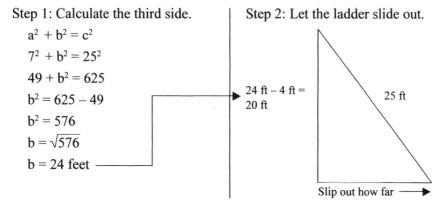

24 ft − 4 ft = 20 ft   25 ft

Slip out how far ⟶

Step 3: Now calculate the third side.

$a^2 + b^2 = c^2$

$20^2 + b^2 = 25^2$

$400 + b^2 = 625$

$b^2 = 625 - 400$

$b^2 = 225$

$b = \sqrt{225}$

$b = 15$ feet

Answer: The ladder slides out 15 feet – 7 feet or 8 feet.

Note: Illustrations are often helpful in the process of calculating an answer. However, I watched one student attack this last question with a picture but he drew it from the wrong angle and never saw that the problem was one of triangles because he drew this:

Notes: ABCD below is a rectangle.

ADE and EFC are 30°–60° right triangles.

AEF is a 45°–45° right triangle.

DE = 1

Angle AEF = 90°

Question: What are the perimeter and area of triangle AFB?

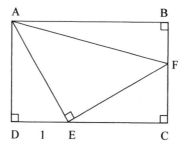

Answer: Step 1: Focus on triangle ADE.

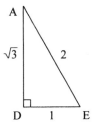

The hypotenuse is twice as long as the short leg in a 30°–60° right triangle. Here is the calculation for the other leg:

$a^2 + b^2 = c^2$

$1^2 + b^2 = 2^2$

$1 + b^2 = 4$

$b^2 = 4 - 1$

$b^2 = 3$

$b = \sqrt{3}$ length of the longer leg

Step 2: Now look at triangle AEF. Using the 30°–60° right triangle properties, I know that $\overline{AE} = \overline{EF} = 2$. Angle AEF = 90°.

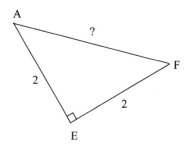

$a^2 + b^2 = c^2$

$2^2 + 2^2 = c^2$

$4 + 4 = c^2$

$8 = c^2$

$\sqrt{8} = c$

$\sqrt{4 \cdot 2} = c$

$2\sqrt{2} = c$ length of hypotenuse $\overline{AF}$

Step 3: Now look at triangle CEF. As in a 30°–60° right triangle, the sides are 1, 2, and √3. $\overline{EC} = \sqrt{3}$.

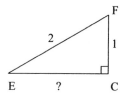

Therefore:
$\overline{DC} = \overline{AB} = 1 + \sqrt{3}$.
$\overline{BC} = \overline{AD} = \sqrt{3}$
$\overline{BF} = \sqrt{3} - 1$

Step 4: Now we can finally draw triangle ABF with dimensions.

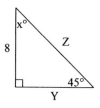

Perimeter of ABF = $2\sqrt{2} + 1 + \sqrt{3} + \sqrt{3} - 1$
$= 2\sqrt{2} + 2\sqrt{3}$ or $2(\sqrt{2} + \sqrt{3})$

Area of ABF = $[(1 + \sqrt{3})(\sqrt{3} - 1)]/2$
$= (\sqrt{3} - 1 + 3 - \sqrt{3})/2$
$= 2/2$
$= 1$ square unit

Example: Find x, Y, and Z in this triangle.

Answer: Since this triangle has a 45° angle and a 90° angle, it must be a 45°–45°–90° special right triangle.

If a right triangle has one 45° angle, the other acute angle must be 45°, so angle x = 45°. The sum of the angles is (90 + 45 + 45 = 180). Correspondingly, if one leg is 8 units, the other is also 8 units. So leg Y = 8 units.

This leaves only the simple calculation for the length of the hypotenuse Z:

$a^2 + b^2 = c^2$

$8^2 + 8^2 = Z^2$

$64 + 64 = Z^2$

$128 = Z^2$

$\sqrt{128} = Z$

$\sqrt{64 \cdot 2} = Z$

$8\sqrt{2} = Z$ or about 11.31

Example: Find sides A, B, and angle c in this triangle.

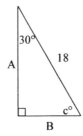

Answers: Step 1: In a right triangle, if one acute angle is 30°, the second is 60° (30 + 60 + 90 = 180). Angle C = 60°.

Step 2: In a 30°–60° right triangle, the short leg is half of the hypotenuse. This means side B = 9.

Step 3: Now to find side A.

$a^2 + b^2 = c^2$

$a^2 + 9^2 = 18^2$

$a^2 + 81 = 324$

$a^2 = 324 - 81$

$a^2 = 243$

$a = \sqrt{243}$

$a = \sqrt{81 \cdot 3}$

$a = 9\sqrt{3}$ or about 15.59

19

Circles

INTRODUCTION

First, let's start with a few definitions and formulae.

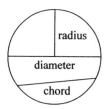

Circle: A closed arc with every point equidistant from the center.

Radius: A line segment whose one endpoint is at the center of the circle and whose other endpoint is on the circle.

Chord: A line segment whose endpoints are on the circle.

Diameter: A special chord that passes through the center of the circle. It is twice the radius (d = 2r) and is the longest chord on the circle.

Pi (π): The ratio of the circumference to the diameter C/D. Its value is approximately 3.14 or $22/7$. This value is a constant and thus will never change.

Area: The number of square units enclosed by the circle ($A = \pi r^2$).

Circumference: The linear distance around a circle ($C = 2\pi r$ or $C = \pi d$). Note: The circumference can never be measured in square units.

RELEVANT CONCEPTS FOR ALL TESTS

Example: What are the circumference and area of a circle with a radius of 15 meters?

Circles 169

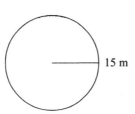

Answer: Circumference = 2πr
C = 2 (3.14)(15)
C = 94.2 meters
Area = πr²
A = (3.14)(15²)
A = (3.14)(225)
A = 706.5 square meters (or m²)

END OF PRAXIS I, 0014, 0511, AND 0146

Now we need to add some additional definitions.

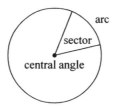

Arc: The distance between two points on a circle.

Semicircle: Half of a circle (180°).

Sector: An area of a circle bounded by an arc and two radii drawn from the center.

Central angle: An angle having its vertex on the center of a circle.

Properties of Circles

The sum of all the central angles equals 360°. Given one angle measure we can find the measure of the other angle.

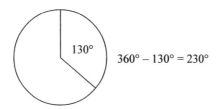

The measure of the central angle in a circle is equal to the measure of the intercepted arc $\overset{\frown}{AB}$.

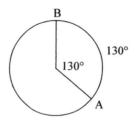

The measure of an *inscribed* angle in a circle is equal to one-half the measure of the intercepted arc $\overset{\frown}{AB}$.

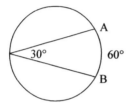

"Inscribed" means that the vertex of the angle lies on the circle.

Example: What is the circumference of a circle whose area is 100π?

Answer:
$A = \pi r^2$

$\dfrac{100\cancel{\pi}}{\cancel{\pi}} = \dfrac{\cancel{\pi} r^2}{\cancel{\pi}}$

$100 = r^2$

$\sqrt{100} = r$

$10 = r$

Therefore $C = 2\pi r$

$C = 2\pi(10)$

$C = 20\pi$

> The formula for the area of a circle is πr^2. Since the area of the circle equals 100π, you can substitute this value into the variable A.
>
> Now divide each side by π to isolate r^2. π cancels on both sides of the equal sign, leaving 100 on the left and r^2 on the right.
>
> r^2 is not in it simplest form. In order to simplify, take the square root of both sides of the equation to find 10 = radius.
>
> *We are not done.*
>
> The question asked for the circumference. Therefore, substitute 10 into the r of the circumference formula to answer the question.

Circles

Example: What is the area of a circle whose circumference is π?

Answer:

$C = 2\pi r$

$\pi = 2\pi r$

$\dfrac{\pi}{\pi} = \dfrac{2\pi r}{\pi}$

$1 = 2r$

$\dfrac{1}{2} = \dfrac{2r}{2}$

$\dfrac{1}{2} = r$

> The formula for the circumference of a circle is $2\pi r$. Since the circumference of the circle equals π, you can substitute this value into the variable C.
> Now divide each side by π to isolate r. π cancels on both sides of the equal sign, leaving 100 on the left and r^2 on the right.
> Now divide by 2 to isolate the r.
> *We are not done.*
> The question asked for the area. Therefore, substitute ½ into the r of the area formula to answer the question.

Therefore:

$A = \pi r^2$

$A = \pi(1/2)^2$

$A = \pi(1/4)$ (note: ½ • ½ • ¼)

$A = \dfrac{\pi}{4}$

Example: What is the area of a circle that is inscribed in a square of area = 2?

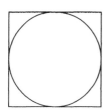

Here a polygon is "circumscribed" about a circle, which means that each side of the polygon is touching the circle at only one point.

Answer: For the square,

$A = S^2$

$2 = S^2$

$\sqrt{2} = S$

Now for the circle, the diameter = $\sqrt{2}$ so the radius = $\sqrt{2}/2$. Thus,

$A = \pi r^2$

$A = \pi(\sqrt{2}/2)^2$

$A = \pi(2/4)$

$A = \pi/2$

Example: In the following figure, what is the perimeter of the portion of the circle bound by the arc AB and the short segment of the circle between A and B? The central angle is 60°.

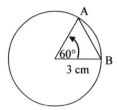

Answer: First consider the triangle. Any triangle with two equal sides and one 60° angle is equilateral. All sides are thus 3 cm long. Side AB is 3 cm long.

Now for the *arc length*: 60° is $^{60}/_{360}$ or $^1/_6$ of the circumference. The formula becomes:

Arc AB = $\dfrac{2\pi r}{6}$

AB = $\dfrac{2\pi 3}{6}$

AB = $\dfrac{6\pi}{6}$

AB = π

> Here we have the circumference divided by $^1/_6$ of the circumference.
>
> Now substitute 3 for the radius (r). Since multiplication is commutative, we can multiply 2 times 3, giving the product of 6.
>
> Now simplify: AB = π.

Finally, we have the two components of the perimeter:

Perimeter = π + 3

Perimeter ≈ 6.14 cm

> This makes perfect sense. If line AB is 3, then arc AB is slightly longer. 3.14 *is* slightly longer.

20

Probability

INTRODUCTION

Probability questions concern themselves with the likelihood of an occurrence. Intertwined with this concept is a related one regarding the number of ways a given outcome could be obtained. The concepts of permutations and combinations are related to the latter.

At its most basic, probability is concerned with a ratio, proportion, or fraction that places successful outcomes over possible outcomes:

Successful outcomes
Possible outcomes

For example, if a given situation is satisfied one out of four times, it would be expressed as 1 to 4, 1:4, or 1/4.

RELEVANT CONCEPTS FOR ALL TESTS

One of the simplest probability questions concerns colored marbles in a bag.

Example: A bag contains 6 red marbles and 8 blue marbles. What is the probability of drawing out one red marble?

Answer: There are 6 possible (red) successes out of 14 possible outcomes. This gives 6/14, or 3/7 (or 3:7), or three chances in seven.

The same problem could be reworded for the probability of *not* pulling out a red marble. Then 8 possible (blue) successes out of 14 possible outcomes, or 8/14, or 4/7 (or 4:7), or four chances in seven.

Slightly harder would be the addition of a third color, say 4 white marbles. Now the probability of selecting a red marble would be 6/18, or 1/3 (or 1:3), or one chance in three.

A question might also ask what would be the probability of pulling out a red marble *or* a white marble. There are 6 red, 4 white, and 8 blue, making a total of 18. Now the probability would be $(6 + 4)/18 = 10/18$, or 5/9 (or 5:9), or five chances in nine.

Traditionally, probability questions have often been couched in terms of playing cards or dice. Let's look at a couple of these.

Example: What is the probability of drawing a heart from a deck of cards?

Answer: Since there are 13 hearts in a deck of 52 cards, the chances are 13 in 52, or 1 in 4 (or 1/4 or 1:4) of drawing a heart.

A twist would be to word it this way:

Example: What is the probability of drawing a heart or an ace?

Answer: There are 13 hearts but one is an ace, so there are only three more aces. This gives 13 + 3 successes of 52, 16/52 (or 4/13) and thus 4 chances in thirteen.

Now let us consider a die (one of a pair of dice). Each has six sides numbered from 1 to 6.

What are the chances of throwing a one? That would be 1 chance in 6 (or 1/6 or 1:6).

What about throwing a 5 *or* a 6? This gives 2 chances in 6 or 1 in 3 (or 1/3 or 1:3).

Slightly more complex might be throwing two ones on two dice. There are 36 possible outcomes (6×6 or 1, 1; 1, 2; 1, 3; 1, 4; 1, 5; 1, 6; etc.), of which only one satisfies the condition (a success). Thus there is only 1 chance in 36 (or 1/36 or 1:36).

What would be the probability of throwing a total of 7 on two dice? There are six possible ways to get a seven: 1, 6; 6, 1; 2, 5; 5, 2; 3, 4; and 4,3. That is 6 ways, 6 out of a possible 36, or 1 chance in 6 (or 1/6 or 1:6).

END OF PRAXIS I, 0014, 0511, AND 0146

At the heart of probability is the concept of randomness. At its most basic level, our view of random is rather different from reality.

Let's create our own example to illustrate this difference. From a class, four students are selected and divided into two pairs. The first pair is assigned the task of writing down a series of 100 calls of "heads" or "tails" *as if* they were flipping a coin. These are recorded like this: H, H, T, H, T, T, H, and so on. The second pair of students is given a coin to flip, and they record the results in the same manner as the first pair. The difference between the groups will be clear. The first pair, the "non-flippers," will have very short groups of repeated Hs or Ts, whereas the second pair recording the actual coin flips will have longer repeats, often 7–10 times! So our *concept* of random is rather different from the actuality.

This same aspect of randomness is also incorporated in what mathematicians call a "random walk." They imagine a pedestrian walking along a road. At each intersection the walker flips a coin. Heads means the pedestrian goes ahead one block, tails means he or she returns a block. Our mistaken impression is that the person will hover a few blocks from the starting point in this random process. Reality is quite different. Chance will show this person may move quite far from the starting point when a "real" random process is utilized. Again, our conception is flawed.

Mathematicians presenting probability love the following illustration (it has even been given a name, the "birthday problem"):

In a group of 50 randomly selected persons, how likely is it that two will share the same birthday?

Most persons initially think that with 366 days (counting leap years) and only 50 people, there is a low probability of duplication, perhaps on the order of about 1 in 7, which is approximately what 50/366 equals. The reality shown by analysis is counterintuitive, with an extremely high probability of a 97% chance of duplication with just 50 people.

An analysis could take this approach. Calculate the likelihood of having *no* matches. Thus the first person has no chance of duplicating, the second has a 365/366 chance of not duplicating, the third has a 364/366 chance, the next has a 363/366 chance, and so on down to the fiftieth person, who has a 317/366 chance of not duplicating. Chances are steadily declining. Since all 50 need to have different birthdays to satisfy the conditions, the likelihood will be small and all fifty fractions must be *multiplied*. The result is actually a small number, .03 or a 3% chance of *not* duplicating, and therefore a 97% chance of duplicating. The turning point actually comes quite early, namely at 23 people. With 23 people the chances are about 50/50 for duplication. If the group size is increased to 90, the probability of duplication reaches .999993 (99.99%), which is a virtual certainty.

Bayes's Theorem

Bayes's Theorem is an important concept when studying *conditional* probability where there are two conditions that need to be satisfied.

Example: What is the probability that a red card is a face card in a deck of 52 cards?

Here are two related restrictions that must be met. The notation for these cases is as follows:

P[A|B] is the probability of A given B

In the case above, that would be the probability of a card being both a face card and red.

Bayes's Theorem: $P[A|B] = \dfrac{P[B|A] \cdot P[A]}{P[B]}$

Example: In a college that has an enrollment of 70% male students and 30% female students, half of the female students (50/50) wear trousers. What is the probability that a student (sex unknown), seen at a distance wearing trousers, is female?

Answer: The data could be summarized in a convenient table where the numbers could be percents or number out of 100 typical students.

	Females	**Males**	**Total**
Trousers	15	70	85
Skirts	15	0	15
Total	30	70	100

P[A] = .30 = probability that a student is female.

P[B] = .85 = probability that any student is wearing trousers (.70 + .15) (all males and trouser-wearing females).

P[B|A] = .5 = probability that a female is wearing trousers (trousers/females = 15/30 = 1/2 = .5).

P[A|B] = ? = probability that a student wearing trousers is female.

$P[A|B] = \dfrac{P[B|A] \cdot P[A]}{P[B]} = \dfrac{.5(.30)}{.85} = .15/.85 = .176 = 17.6\%$

The Importance of the Words "and" and "or" in Probability Questions

In general, the use of "or" in a condition denotes that the probabilities are *added*, but "and" indicates that the probabilities are multiplied. This is logical because when there are alternative ways (addition) to satisfy a condition, there is a higher likelihood of it happening. In the same way, when there are more restrictions (for example, when one event must be followed by another), multiplication is required and the probability will decrease. Here are some examples:

1. What are the probabilities of drawing a jack or a queen from a deck of cards?

   Probability of drawing a jack: 4/52 = 1/13

   Probability of drawing a queen: 4/52 = 1/13

   Probability of drawing either: (4 + 4)/52 = 8/52 = 2/13

   The probability would increase if two draws were allowed:

   2/13 + 2/13 = 4/13

2. But that changes if a jack *and then* a queen must be drawn *in order* (in two draws). (First card replaced.)

   Probability of drawing a jack: 1/13

   Probability of drawing a queen: 1/13

   Probability of both occurring: 1/13 × 1/13 = 1/169

With or without Replacement

In some cases the first choice changes the odds for the second case (dependent events).

Example: What are the chances of drawing two tens from a deck of cards? (A ten *and* a ten without replacement.)

Chances of drawing a ten as the first card: 4/52.

Chances of drawing a ten as the second card: 3/51.

Chances of both happening: 4/52 × 3/51 = 12/2,652 = 1/221.

Combinations and Permutations

Two definitions are needed here:

Permutations are all the different ways selected items may be arranged. Order is a factor: that is, AB and BA are different and count as two occurrences.

Formula: $_nP_r = \dfrac{N!}{(N-R)!}$

(Read as permutations of n things taken r at a time) (! = factorial, thus 5! = 5 × 4 × 3 × 2 × 1 = 120)

Example: What are the permutations of 20 things taken 2 at a time?

Answer: $20!/(20-2)! = 20!/18! = 20 \times 19 \times 18 \times 17.../18 \times 17 \times 16... = 20 \times 19 = 380$

Combinations are the same as permutations above, but order is *not important* and thus AB and BA are counted as one occurrence.

Formula: $_nC_r = \dfrac{N!}{R!(N-R)!}$

And so the combination of twenty things taken two at a time ($_{20}C_2$) would be:

$20!/2!(20-2)! = (20 \times 19 \times 18 \times 17...) / 2(18 \times 17 \times 16 \times 15 \times 14...)$

This simplifies to: $(20 \times 19) / 2 = 10 \times 19 = 190$. This is exactly half of the permutation example, as it should be.

Permutation word problem: A teacher has twelve student papers for display at a PTA open house. She only has room for four at a time. How many different ways can she display them?

$_{12}P_4 = 12! / (12-4)! = (12 \times 11 \times 10 \times 9...) / (8 \times 7 \times 6 \times 5 \times 4...)$
$= 12 \times 11 \times 10 \times 9 = 11,880$

Combination word problem: Same conditions as above but order does not matter. That is, how many different ways may four be chosen?

$_{12}C_4 = 12!/4!(12-4)! = (12 \times 11 \times 10 \times 9 \times 8 \times 7...) / (4 \times 3 \times 2)(8 \times 7 \times 6 \times 5...)$
$= (12 \times 11 \times 10 \times 9) / (4 \times 3 \times 2) = 11 \times 5 \times 9 = 495$ different ways

21

Decoding Tables and Graphs

INTRODUCTION

One of the main goals of test makers is to assess the understanding of test takers. Test makers are usually not interested in measuring a person's level of knowledge. Remember, understanding is not knowing. To do this in math, test makers often rely on the interpretation of tables and graphs. This approach is effective because no student can "learn" all the different graphs or tables that could be presented; rather, they must understand each separate presentation.

Unsuccessful students often concede that they simply guess on table or graph questions and move on because they have no mental tools to decode the tables and graphs.

RELEVANT CONCEPTS FOR ALL TESTS

The first step for successful interpretation and decoding of tables and graphs is to have a plan. We suggest the following:

- Title
- Caption, description, or source
- Horizontal scale
- Vertical scale
- Actual data, information, statistics, bars, lines, points, etc.

On each of the above steps the testee must understand the test item and not just "look at" the information.

A common method of presentation is for a test to contain a visual (table or graph) followed by several questions. A productive approach is to disregard the questions and focus first on decoding the graphic. Once the table or graph is decoded, focus on the individual test items. In other words, invest in decoding before attempting the answers to the questions. To ensure that that process is followed, a test taker can even cover the questions with his or her hand until the decoding process is complete. This removes the possibility of "cheating" and going to the questions first.

A common, but relatively unsuccessful, approach is to read a question and then look up to the graphic to locate the answer. Students using this approach try to avoid getting a general understanding and hope that they can just "find" an answer. This approach is analogous to students on a reading test who are asked an inferred main idea or a "best title for this selection" question. Since the passages may be quite long and intimidating for a poor reader, these students simply scan the passage looking for the answer. They treat an inferential comprehension question just as they do a literal comprehension one. This is a fundamental error. However, since main ideas are not generally stated, students miss almost all questions answered with this approach. On a test like the PRAXIS I Math, which contains almost one-third of the questions as graphics, this method is a sure recipe for failure.

Quickly survey the table below before answering the following questions.

European Union Election Results for 1999 and 2004, by Party

Group	Seats (1999)	Seats (2004)
EUL	49	39
PES	210	200
EFA	56	42
EDD	19	15
ELDR	60	67
EPP	272	276
UEN	36	27
Other	29	66
Total	731	732

1. *What do we know about this table?* It is a table that is *comparing* the number of delegates that occupied the seat for their political party from two specific years (1999 and 2004).
2. *Find the percent increase or decrease*, which means finding the difference between both totals, then dividing the difference by the original number. This will give either the percent increase or decrease. For example, what is the percent decrease for PES between the two years? $210 - 200 = 10$, and $10/210 = .0476$, or a decrease of about 5 percent.
3. *Find the percent of one or more categories.* For example, the percent of EDD in 2004 is 15/732, or approximately 2 percent.

Decoding Tables and Graphs

4. *What is the ratio of EFA to ELDR for 1999?* 56:60.

5. *What is the ratio of EFA to ELDR for 2004?* 42:67.

6. Note: The "Other" category is not one party but a grouping of a number of small parties.

Now we are ready to answer the following questions:

Example: In 1999 the two largest parties comprised what percent of the total number of delegates?

Answer: The two largest parties, PES and EPP, had 482 delegates (210 + 272). The total number of delegates was 731. So 482/731 is .66, or 66 percent.

Example: In 1999, what percent was comprised by the two smallest parties?

Answer: The total was 48 delegates (19 + 36). So 55/731 is .075, or 7.5 percent.

Example: What are the corresponding numbers for 2004? (Two largest and two smallest.)

Answer: The two largest are 200 + 276 = 476 of 732 delegates, or 476/732 = .65, or 65 percent. The two smallest are 15 + 27 = 42 of 732 delegates, or 42/732 = .057, or 5.7 percent.

Quickly survey the graph below before answering the following questions.

Heights of Black Cherry Trees

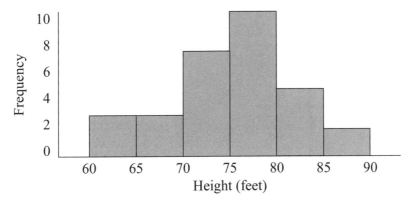

1. *What do we know about this graph?* The graph, which depicts the heights of trees, is an absolute frequency graph, which groups information into intervals that may or may not include the end points on the x axis. For example, the first interval is 60 to 65. Although the end points are 60 and 65, it does not guarantee that there are trees that have a height of 60 or 65

feet. And by extension we can't even tell the exact height of any of the trees within that interval, other than that the trees are between 60 and 65. This holds true for all of the intervals in a relative frequency graph; therefore, approximation is used.

As the graph moves to the right on the x axis the heights increase, and as the graph moves up the y axis the frequency increases. From this information, we can conclude that there are more trees between the interval height 75 to 80; however, we cannot distinguish how many trees have the height of 75, if any, or 76, 77, and so on. This means that we are talking about questions involving interval characteristics. Therefore determining a mean, median, or mode is not so straightforward.

We know that although the y axis interval ends with a 10, it does not mean that there are only 10 trees. In relative frequency graphs, the y and x axes work together. Looking at the x axis only gives partial information, and looking at the y axis only gives partial information. It takes both axes to draw a conclusion about the data. Here, since the y axis is not in intervals of percents but in intervals of quantities, we can figure out the total number of trees that are being considered. So how many trees are we talking about here? Well, there are 6 trees between 60 and 70, there are 7 trees between 70 and 75, there 10 trees between 75 and 80, and so on, totaling 30 trees. This is a simple concept, but many times misunderstood, and reveals your understanding about how to interpret graphs.

2. *What do the intervals tell us?* The interval between 70 and 80 has a total of 18 trees, which is a little more than half the total number of trees. The interval that exceeds 80 includes 7 trees, which is a little less than a quarter of the trees. The interval that is less than 70 includes 6 trees, which is about a fifth of the trees.

Now we are ready to answer the following questions:

Example: How many trees have a height of less than 75 feet?

Answer: $3 + 3 + 7 = 13$ trees.

Example: How many trees are represented in this graph?

Answer: $3 + 3 + 7 + 10 + 5 + 2 = 30$

Example: What percent of the trees exceed 80 feet in height?

Answer: $5 + 2 = 7$ trees. The total number of trees is 30, so $7/30 = .233$, or 23.3 percent.

Quickly survey the graph on the next page before answering the following question.

Travel Time for Commuters

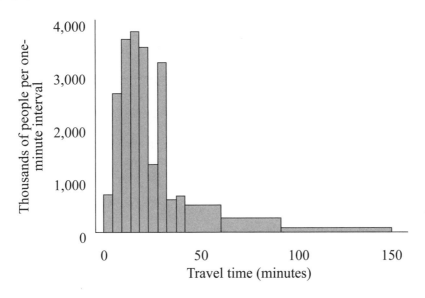

1. *What do we know about this graph?* It is an absolute frequency graph that displays the travel times of commuters. The x axis is in intervals of 50 minutes and the y axis is the number of people in intervals of 1,000. However, the side heading for the y axis is important. Notice that it states "thousands" of people per one minute. How should this be interpreted? Should it be ignored because the y axis is already in intervals of 1,000? No, because this side heading is saying to multiply 1,000 by 1,000, 2,000, 3,000, and so on to give the accurate number of people it represents. The side heading is a matter of convenience to the interpreter. It would have been just as easy to have the interval on the y axis in millions.

   We can see that most of the commuters travel less than an hour. We can also see that there are few people who have a longer commute, which tells us that our graph is skewed. So if we are asked questions about the central tendencies (mean, median, and mode), the mean would not be a good indicator. Instead, the median would be the best tool to use to describe the data. The mode is easy; it is the tallest bar.

Now we are ready to answer the following questions:

Example: The modal commute travel time is about how many minutes?

Answer: The tallest bar is at about 20 minutes.

Example: About how many people are represented by that bar?

Answer: The scale is in thousands, so the first number 1,000 represents 1 million. This gives the tallest as almost 4 million.

Make sure that each side heading is considered. Here it would be easy to think that each interval on the vertical axis is in thousands; however, if you read the side heading you will know that each interval is then multiplied by 1,000, giving the true representation of how many people there are for each interval of time.

Quickly survey the graph below before answering the following question.

English Dialects in 1997

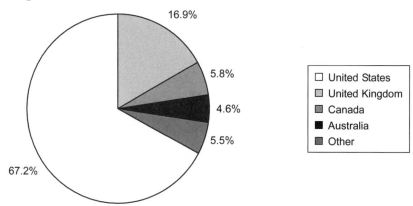

1. *What do we know about this graph?* It is a pie graph used to compare various items. In this case we are comparing the percentage of English dialects in 1997. This means that questions could be in terms of fractions or decimals.

    The United States makes up approximately two-thirds of the pie. The United Kingdom and the rest of the countries make up approximately one-third of the pie. Canada, Australia, or Other are each approximately one-third of the United Kingdom's total. The United Kingdom's total is approximately one-fourth of the U.S. total.

    If the total of the number of people in any one area is given, we will be able to tell how many people in the other areas speak English using proportions.

Now we are ready to answer the following questions:

Example: If 280,000,000 people speak English in the United States, how many speak English in the United Kingdom?

Answer: This is best approached as a proportion. Set it up this way:

$$\frac{280,000,000}{x} = \frac{.672}{.169}$$

.672x = 280,000,000(.169)

.672x = 47,320,000

x = 47,320,000/.672

x = 70,416,666

Quickly survey the table below before answering the following questions.

Distance traveled from St. Louis to Sacramento

Time in hours	Distance from St. Louis (miles)
1	60
2	120
3	180
4	240

(differences of 60 between consecutive distances)

1. *What do we know about this table?* This is a distance, rate, and time problem. Looking at the pattern as a function of time, miles increase by 60 each hour from the starting point in St. Louis. Given this table, we can conclude three things: the time of the trip, the distance traveled after each hour, and the constant rate of 60 miles per hour, which means that the function is linear with a positive slope of 60.

   The equation will be y = 60x, rewritten as d = 60t.

Total time: 1,800 = 60t

$$\frac{1{,}800}{60} = \frac{60t}{60}$$

30 = t   Total travel time from St. Louis to Sacramento

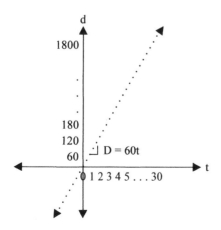

Notice that the linear graph goes through the origin, which reflects the table above. At zero time, distance will also be zero. The graph is just another way to describe the relationship between time (t) and distance (d) from St. Louis to Sacramento. Notice that the equation of the line is d = 60t (going up from left to right), indicating a positive slope.

Now we are ready to answer the following question:

A train moving at a constant speed leaves St. Louis for Sacramento at time = 0. If Sacramento is 1,800 miles from St. Louis, which equation describes the distance from Sacramento at time = t?

1. d = 60t – 1,800
2. d = 60t
3. d = 1,800 – 60t
4. d = 1,800 + 60t

Be careful! The question is *not* asking what we anticipated from the table. If we did not take the time to read the question stem, we would have given the wrong answer (choice 2) because we assumed the question would come directly from the table. The question is asking, "what is the *remaining distance* to Sacramento?" at any given time. This question is not so direct; therefore, we must change the approach. What information can we use? The constant rate of 60 is still good but the equation must be modified to have a negative slope instead of a positive slope. Notice that the question gives the total distance of 1,800 miles. This clue should alert you that either the question wants the total time it will take to get to Sacramento, which will fit the table, or the distance remaining at any given time, which will not fit the table. Here it clearly states that it wants an equation that depicts the distance from Sacramento at "t" time. Now, with this new approach the modified table below will clarify how to answer the question visually.

Time in hours	Distance from St. Louis (miles)
0	1,800
1	1,740
2	1,680
3	1,620

Notice that this table is the opposite of the one on the preceding page. It indicates that the graph will have a y intercept at 1,800 and a negative slope of 60, which reflects the question that is being asked. You can see this on the modified table by noticing that as time increases, distance decreases, and where it reads zero time is where the linear graph crosses the y axis. Compare the graph on the following page with the table above.

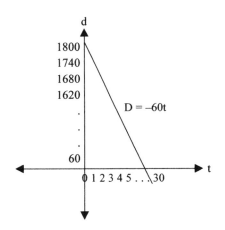

The table would reach 30 hours at zero distance. Notice also that the equation of the line (D = –60t) is going down from left to right, indicating a negative slope.

Answer: The basic formula is d = rt but the question wants to know the constantly decreasing distance from Sacramento, *not* the distance from the starting point; therefore, the answer is number 3.

D = 1,800 – rt

D = 1,800 – 60t

Example: As a project, your class has collected recyclables to finance a class trip. Your aim is to maximize the income. You decide to sell each separate category to the recycler who pays the most. How much will you get for the following?

Aluminum	2,500 lbs.
Cardboard	1,300 lbs.
Glass	3,700 lbs.
Plastic	1,100 lbs.

Recycler	Aluminum	Cardboard	Glass	Plastic
A	.30/lb.	.03/lb.	.08/lb.	.02/lb.
B	.35/lb.	.04/lb.	.07/lb.	.03/lb.

Answer: Your income will be:

Aluminum (2,500)(.35) = $875

Cardboard (1,300)(.04) = $52

Glass (3,700)(.08) = $296

Plastic (1,100)(.03) = $33

Total = $1,256

END OF PRAXIS I, 0014, 0511, AND 0146

It is common on these tests to be asked for calculations of *linear equations* that involve finding the slope. What is slope? It is a numerical measure of the steepness of a line that is often referred to as the ratio of rise over run.

Slope: Given two points in the format (x, y), the slope (m) is determined by this formula:

$$\frac{y_2 - y_1}{x_2 - x_1} = m$$

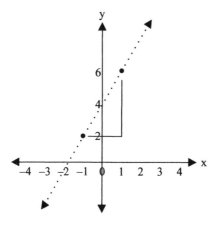

$$\frac{y_2 - y_1}{x_2 - x_1} = m$$

$$\frac{6 - 1}{2 - (-1)} = \frac{5}{3}$$

x	y	
-2	0	(-2, 0)
-1	2	(-1, 2)
0	4	(0, 4)
1	6	(1, 6)
2	8	(2, 8)
3	10	(3, 10)

The x and y values are ordered pairs and coordinates are points on the graph above.

The point where x and y meet is called a coordinate pair (x, y). As you can see, there is an infinite number of points on a line. To determine the slope, just find any two points on the line and substitute them into the slope formula. It does not matter which y value is assigned to y_1 or y_2, or which x value is assigned to x_1 or x_2, as long as you are consistent. For example, on the left we arbitrarily picked two points on the line, which are (-1, 2) and (1, 6). The first number in a coordinate pair is always the x value and the second number is always the y value. We chose to substitute the 6 in for y_1; therefore 1 in the first coordinate pair will have to be assigned to y_2. To keep consistency with the assignment of the y values, the x values will follow the same order, which means 2 in the first coordinate pair will be assigned x_1 and -1 will have to be assigned x_2 (see the calculation on the left). Notice that the y value will always be the numerator and the x value will always be the denominator. The slope is 3 units to the right on the x axis and 5 units up the y axis. This same pattern will continue throughout the line no matter how far the line is extended (see the table on the left). Based on the pattern of the table, it is easy to conclude that other (x, y) values could be (2, 8) and (3, 10).

Example: If a linear equation passes through (0, 2) and (2/3, 0), what is the slope?

$$\frac{y_1 - y_2}{x_1 - x_2} = m$$

$$\frac{0 - 2}{(2/3) - 0} = \frac{-2}{(2/3)} = -2(3/2) = -3$$

(See Chapter 11 for a review on dividing complex fractions.)

Let's see how this slope is graphed.

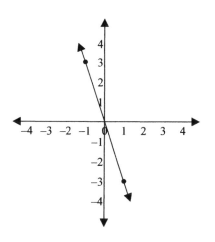

Notice two things:

1. The slope is negative; therefore, the graph will go down from left to right.

2. The slope is not a fraction like the one in the previous example. In the previous example the slope 5/3 is really the y units over the x unit of 1, which means that as x changes one unit, y changes 5/3 units upward. Why didn't we divide 5/3 by one? Because a complex fraction that is divided by a fraction in the form of one (1/1) will remain the same. Here –3 is not a complex fraction but it still represents –3 y units. In this case, Y falls –3 units as x runs 1 unit (–3/1). This subtle concept can be misleading if you do not understand what is really taking place with the relationship x and y. So if the question is asked, "How many units does x change to the y units?" it will be one unit.

The two ways linear equations can be written are the *standard form* and the *slope-intercept form*. The standard form is in the form of ax + by + c where a, b, and c are constants.

```
      variables
       ↑  ↑
    ax + by + c = 0
       ↘  ↓  ↙
       constants
```

Let's look at an example to explain how to graph a linear equation in standard form and understand what happens to the equation when one or more of the constants is equal to zero.

$$\underset{a}{9x} + \underset{b}{3y} + \underset{c}{6} = 0.$$

To make this linear equation easier to work, rewrite $9x + 3y + 6 = 0$ to:

$$\underset{a}{9x} + \underset{b}{3y} = \underset{c}{-6}$$

This is still considered standard form where a = 9, b = 3, and c = –6.

This equation can be easily graphed by substituting 0 for x to solve for the y value and then substituting 0 for y to solve for the x value, but not at the same time or the equation will result in no solution ($0 \neq -6$).

$9(0) + 3y = -6$ $9x + 3(0) = -6$
$\quad\quad 3y = -6$ $\quad\quad 9x = -6$
$\quad\quad\;\; y = -2$ $\quad\quad\;\; x = -2/3$

$(0, -2)$ $(-2/3, 0)$
ordered pair where   ordered pair where
x = 0 and y = –2   x = –2/3 and y = 0

Now, plot the ordered pairs and draw a line from one point to the other to see geometrically the standard form of the equation $9x + 3y = -6$. The slope of the line is not so easy to see when the equation is written in standard form; therefore, the slope formula must be used to find the slope.

$$\frac{y_1 - y_2}{x_1 - x_2} \rightarrow \frac{0 - -2}{-2/3 - 0} = \frac{2}{-2/3} = -3 \text{ is the slope of the line.}$$

The slope-intercept form is the preferred method of a written linear equation because the slope and the y intercept are so easily identifiable. See how the equation is written:

Decoding Tables and Graphs

The constant is where the line crosses the y axis—the y intercept.

$y = ax + b$ or $y = mx + b$ (preferred method)

The slope is beside the x variable.

It is easy to transform a linear equation in standard form to a slope-intercept form by simply isolating the y variable. Let's demonstrate by using the same equation ($9x + 3y = -6$) that was used in the previous example.

$9x + 3y + 6 = 0$
$ -6 -6$
$9x + 3y = -6$
$-9x -9x$
$\dfrac{3y}{3} = \dfrac{-9x}{3} - \dfrac{6}{3}$
$y = -3x - 2$

1. Subtract the constant 6 from both sides of the equals sign using the additive inverse property.
2. Subtract 9x from both sides of the equals sign, again using the additive inverse property.
3. Now divide each term by 3 using the multiplicative inverse property to isolate the y variable.
4. Notice that the slope is the same as above and easily identified, which alleviates the step of solving for the slope; however, the x intercept is not so readily seen.

To solve for the x intercept when the equation is in slope-intercept form, substitute 0 into the y variable and then solve for x.

$y = -3x - 2$
$(0) = -3x - 2$
$+2 +2$
$\dfrac{2}{-3} = \dfrac{-3x}{-3}$
$-^2/_3 = x$ The x intercept is -0.67.

So both $9x + 3y + 6 = 0$ and $y = -3x - 2$ are equlivalent linear equations and represent the same line.

This checks, because the x intercept is $-2/3$, the y intercept is -2, and the slope is -3.

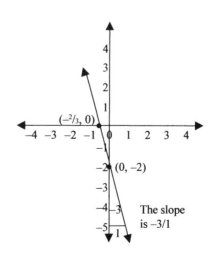

The slope is $-3/1$

Now let's see what happens when the slope or the constant is set to zero in the slope-intercept form equation.

$y = ax + b$
$y = (0)x + b$
$y = b$

If the slope (a) is zero then it will be a horizontal line through the y axis. This means there is no slope. The equation of the line will be either $y = b$ or $y = -b$, where b is a constant.

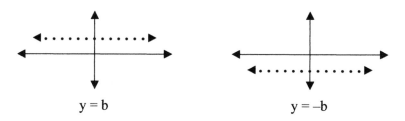

$y = ax + b$
$y = ax + (0)$
$y = ax$

If the constant (b) is zero and a is one, then the graph will be a slanted line through the origin. If a is positive or negative one, then $y = x$ and are said to have symmetry about the origin.

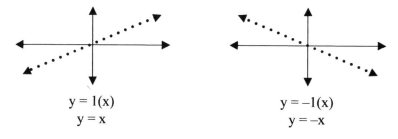

What happens when the line is vertical?

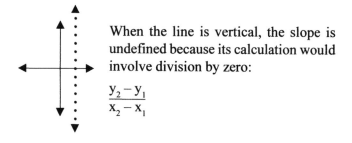

When the line is vertical, the slope is undefined because its calculation would involve division by zero:

$$\frac{y_2 - y_1}{x_2 - x_1}$$

Now let's see what happens to the line as a increases or decreases.

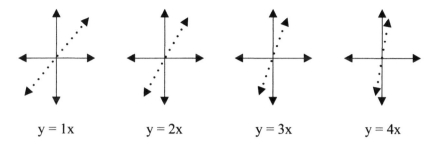

As a increases, the line gets steeper or more vertical.

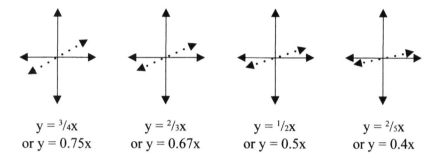

As a decreases, the line gets flatter or more horizontal.

GRAPHING SYSTEMS OF EQUATIONS

When graphing systems of linear equations there are ways to identify whether the systems are parallel, perpendicular, or coincide just by looking at their slopes and y intercepts when the equations are written in slope-intercept form. Being able to identify these components will help save time during testing.

If the slopes of two equations are equal, with different constants (b), the slopes are *parallel*.

$y = 3x + 2$
$y = 3x - 2$

If the product of two slopes is –1, the lines are *perpendicular*.

y = –x + 2
y = x – 2
(–1)(1) = –1

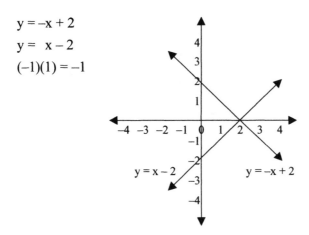

y = x – 2   y = –x + 2

Not reflected in this graph, other slope pairs indicating perpendicular lines could be:
(–2, ½)
(–3, ⅓)
(4, –¼)
(5, –⅕) etc.

Distance between Two Points

Often you may be asked to find the distance between two points, each presented in the (x, y) format.

$D = \sqrt{(x_2 - x_1)^2 + (y_2 - y_1)^2}$ $\left\{\begin{array}{l}\text{Notice that this equation is a Pythagorean} \\ \text{Theorem adaptation.}\end{array}\right.$

Example: What is the distance between (3, 6) and (7, 9)?

Answer: $\sqrt{(7-3)^2 + (9-6)^2}$
$\sqrt{4^2 + 3^2}$
$\sqrt{16 + 9}$
$\sqrt{25}$
5

The distance between the two points (3, 6) and (7, 9) is 5 units.

Decoding Tables and Graphs

It is easy to measure vertical and horizontal distances but not a slanted distance. The purpose of the distance formula is to measure the length of a slanted line by an alternative of the Pythagorean Theorem ($a^2 + b^2 = c^2$).

Notice that two points are given but not the equation of the line. By looking at the graph you can determine the equation of the line.

1. The slope is positive because the line is going upward from left to right.
2. The line crosses the y axis at approximately $3\frac{1}{2}$. The y intercept is calculated using the slope formula with one point (7, 9) and the other (0, ?). See below.
3. Form a right angle from the given points. This will give the slope of the line by counting the number of units up the y axis (3) and the number of units out the x axis (4) then substitute into y/x. These values makes sense according to Pythagorean Triples (e.g., 3, 4, 5).
4. Based on the information taken from the graph, the equation of the line should be $y = \frac{3}{4}x + \frac{15}{4}$.
5. The line of the equation can be found algebracally by using the slope and the point-slope formulas. Let's check.

Slope Formula

$$\frac{y_2 - y_1}{x_2 - x_1} \rightarrow \frac{9-6}{7-3} = \frac{3}{4}$$

Finding the Intercept

$$\frac{9-b}{7-0} = \frac{3}{4}$$

$$9 - b = \frac{21}{4}$$

$$9 - \frac{21}{4} = b$$

$$\frac{15}{4} = b$$

Point-Slope Formula

$$y - y_1 = m(x - x_1)$$
$$y - 6 = \frac{3}{4}(x - 3)$$
$$y - 6 = \frac{3}{4}x - \frac{9}{4}$$
$$\underline{+6 \qquad\qquad +6}$$
$$y = \frac{3}{4}x + \frac{15}{4} \qquad \text{It's a match!}$$

Line Midpoint

Similarly, a common question asks for the midpoint of a line given two endpoints in the (x, y) format. These simply average the x and y values.

x value of the midpoint $= \dfrac{x_1 + x_2}{2}$

y value of the midpoint $= \dfrac{y_1 + y_2}{2}$

Example: What is the midpoint of the line segment between (–2, 3) and (4, 0)?

Answer: $x = \dfrac{-2+4}{2} = \dfrac{+2}{2} = +1 \qquad y = \dfrac{3+0}{2} = \dfrac{3}{2} = 1.5$

The midpoint is (+1, +1.5).

Example: In the table below, fill in the missing amounts. Calculate the percents for the "Actually Did" row. Construct a circle graph of that data. Label all work.

Student Educational Intent after High School

	No More School	Two-Year Tech School	Four-Year College	Totals
Actually Did	510		375	1,020
Did Not	110			
Total			400	1,200

Answer: First fill in the missing numbers.

	No More School	Two-Year Tech School	Four-Year College	Totals
Actually Did	510	135	375	1,020
Did Not	110	45	25	180
Total	620	180	400	1,200

Analysis of "Actually Did" row:

510 (No more school):
$^{510}/_{1020} = .50 = 50\%$

First, find the percent by dividing the total of "Actually Did" into the part "No More School."

$.50 \cdot 360° = 180°$

Now find the sector of the circle that represents 50% by multiplying the decimal form of 50% by 360 degrees.

135 (two-year tech school):
$^{135}/_{1020} = .13 = 13\%$

Second, find the percent by dividing the total of "Actually Did" into the part "Two-Year Tech School."

$.13 \cdot 360° = 46.8°$

Now find the sector of the circle that represents 13% by multiplying the decimal form of 13% by 360 degrees.

375 (four-year college):
$^{375}/_{1020} = .37 = 37\%$

$.37 \cdot 360° = 133.2°$

Last, find the percent by dividing the total of "Actually Did" into the part "Four-Year College."

Now find the sector of the circle that represents 37% by multiplying the decimal form of 37% by 360 degrees.

**What Students Actually Did after High School:
Results of a Phone Survey of 1,200 Students**

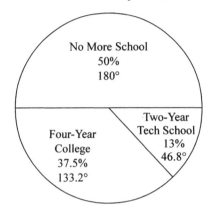

Example: What is the slope of this equation?
  $3x - 4y = 7$

Answer: The formula can be converted to the $y = mx + b$ format:

$3x - 4y = 7$

$\quad -4y = 7 - 3x$

$\quad \dfrac{-4y}{-4} = \dfrac{7}{-4} - \dfrac{3x}{-4}$

$\quad y = {}^{3x}/_4 - {}^7/_4$

$m = {}^3/_4$, which is the slope (for every increase of 1 on x, y will increase ¾ of a point).

Confirmation using the slope formula:

$(y_1 - y_2)/(x_1 - x_2) = m$ (the slope)

$(-1^3/_4 - 0)/(0 - 2^1/_3) = m$

$(-^7/_4)/(^7/_3) = m$

$(^7/_4)/(^7/_3) = m$

$(^7/_4)(^3/_7) = m$

$^3/_4 = m$

Make a table of values and draw:

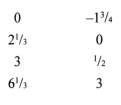
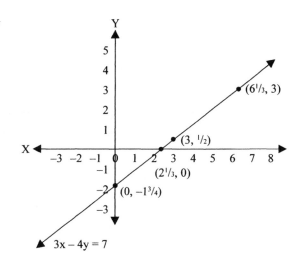

Values of x	Values of y
0	$-1^3/_4$
$2^1/_3$	0
3	$^1/_2$
$6^1/_3$	3

Example: Which of the following might be true?

a. Values of x in equation A are negative.

b. Values of x in equation B are negative.

c. The slope of the equation perpendicular to equation B is –1.

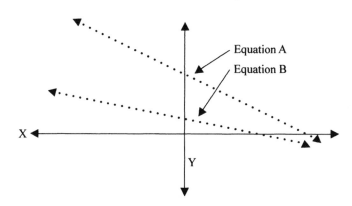

Which would be true?

1. a and c
2. a and b
3. b and c
4. a, b, and c
5. c

Answer: By analysis, some values of x for Equation A could be negative in quadrant II. The same would be true for Equation B in the same quadrant. However, since equation B has a *negative* slope, its perpendicular has to be positive (i.e., its product must be −1). This means that a and b are true, but c is not, so item 2 is the correct answer.

Example: The relationship between the number of sides of a polygon and the sum of the interior angles is as follows: triangle = 180°, quadrilateral = 360°, pentagon = 540°, hexagon = 720°.

Express the relationship in a formula. Draw a graph of the relationship with the number of sides on the x axis and the number of degrees on the y axis.

Answer: The formula would be:

degrees = (n − 2)180° (Where n = number of sides)

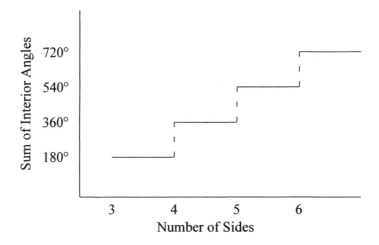

Relationship between the Number of Sides of a Polygon and the Sum of the Interior Angles

Linear equations (lines) are the simplest types of graphs that can be drawn; however, this is not where your understanding should stop because in a real-world curves are everywhere, and we need to understand how they behave. We can do this by using models that are in the form of graphs to observe patterns. One type of graph that forms a curve is called a parabola. Its equation in its simplest form contains x^2 or a^2, both used interchangeably. Its equation in its complete form is $ax^2 + bx + c = y$, where a does not equal zero. Here is an example of a quadratic equation in its simplest form: $y = x^2$.

The graph below has one curve that is symmetrical to the y axis. Each point on this curve will have an x value and a y value; these values are known as a coordinate pair. The y value will be the same on either side of the graph; however, the x values will have opposite signs.

Notice that when a negative or positive x value is substituted into the function x^2, the product is the same. Why? Remember that in Chapter 7 we explained that a negative times a negative gives a positive. The tables below confirm the points on the graph, making a graph in the shape of a parabola.

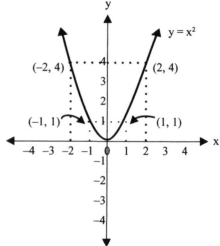

x	x^2	y	and	x	x^2	y
0	0^2	0		0	0^2	0
−1	$−1^2$	1		1	1^2	1
−2	$−2^2$	4		2	2^2	4
−3	$−3^2$	9		3	3^2	9

Now, if the function was $-x^2$, how will the graph behave and what will be the output (y values)? The parabola reflects across the x axis and the y values are no longer positive. Why? When a negative is placed in front of a variable without a written coefficient, it is understood to be one. So actually the $-x^2$ is the same as $-1x^2$; therefore, the x value is substitued into the x, then squared, then multiplied by negative 1—nothing more than the order of operations.

x	$-x^2$	y	or	x	$-x^2$	y
0	$(0)^2$	= 0		0	$(0)^2$	= 0
−1	$(-1)(-1^2)$	= −1		1	$(-1)(1^2)$	= −1
−2	$(-1)(-2^2)$	= −4		2	$(-1)(2^2)$	= −4
−3	$(-1)(-3^2)$	= −9		3	$(-1)(3^2)$	= −9
	↑				↑	

Following the order of operations, -1^2 or 1^2 equals a positive 1, then it is multiplied by a negative 1 making the y value negative. This will be true for all the x values substituted into $-x^2$; however, the y values will still be the same whether the x value is negative or positive. This is reflected on both tables and on the graph.

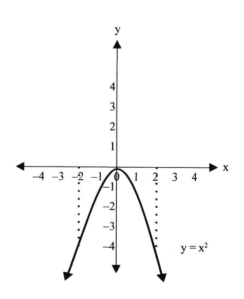

What happens to the parabola when the coefficient of x^2 increases or decreases? If the coefficient increases, it will stretch to become more narrow. If the coefficient decreases, the parabola will become wider. See the examples below.

As the coefficient of x^2 increases the parabolas become narrower. We used 1 and -1 as fixed x values to illustrate that when they are substituted into the x^2 variable, then multiplied by an increasing coefficient, the y values will align themselves, creating a vertical line going through each point where the coordinates meet. Notice also that as the coefficient decreases the parabola becomes wider.

How should this information be interpreted? The narrower the parabola, the faster x increases; the wider the parabola, the slower x increases. A type of real-world situation like acceleration would fit. The orientation of the parabola upward or downward depends on the sign.

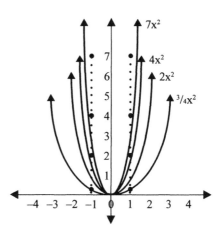

What if you are given a quadractic equation that is in the form $ax^2 + bx = y$? How will it graph? Let's see using the example $x^2 + 4x = y$ and $x^2 - 4x = y$.

When there are only two terms that consist of x^2 and x but not c, then it is understood that the c in the quadratic equation $ax^2 + bx + c = y$ is equal to zero. Look at the first graph. Notice that the parabola has shifted horizontally to the left. Also notice that the vertex of the parabola is directly under the -2 on the x axis and the right side of the parabola crosses the origin (0, 0). The bx is what determines the direction of the parabola going left or right. When the bx value is positive it shifts to the left and when it is negative it shifts to the right (see the second graph). It does the opposite of what you expect it to do. The c is where the parabola crosses the y axis. Since there is no c value it is understood to be zero, just as the graph reflects.

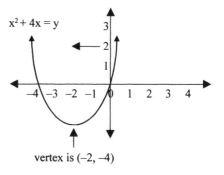

vertex is (−2, −4)

Let's see what happens when the constant (c) is equal to 2. It crosses the y axis at 2. Although the graphs model positive x^2 parabolas, the same pattern applies to negative x^2 parabolas. The only difference is the direction the parabola will open.

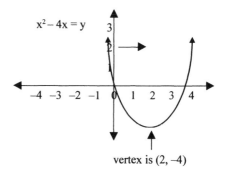

vertex is (2, −4)

Now, how did we know where to align the vertex? Remember that the second term determines the position of the parabola going from left or right horizontally on the x axis. In this case, the 4x is the term that will be used. The procedure is to divide the second term by 2. Why? Remember that a parabola has symmetry if it is folded in half. This symmetry gives two x values that are symmetrical to each other. So when the second term is divided by 2, it divides the area inside the parabola equally.

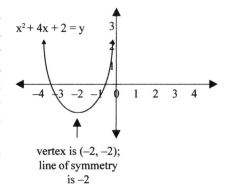

vertex is (–2, –2);
line of symmetry is –2

If the equation had a coefficient greater than 1 in the first term, the procedure would be to divide the coefficient through each term before dividing the second term to locate the vertex. For example:

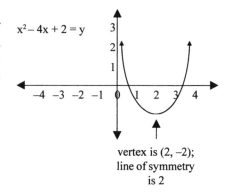

vertex is (2, –2);
line of symmetry is 2

$2x^2 - 4x + 2 = y$

$2x^2 - 4x + 2 = y$ The first term is greater than 1.

$\dfrac{2x^2}{2} - \dfrac{4x}{2} + \dfrac{2}{2} = \dfrac{y}{2}$ Divide the coefficient through each term to simplify the equation.

$x^2 - 2x + 1 = y/2$

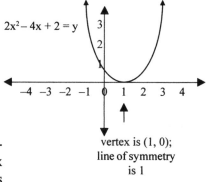

vertex is (1, 0);
line of symmetry is 1

Now divide the second term by 2 to locate the vertex. 2x divided by 2 equals x.

Since the second term is negative, the parabola will shift to the right (positive 1 on the x axis) and the left side of the parabola will cross the y axis at positive one. So looking at the equation, I can visualize the approximate location of the graph just by knowing how each term behaves.

Another way to think about the direction of a parabola is that when the parabola is open upward, its vertex is at its minimum or lowest point, and when the parabola is open doward, its vertex is at its maximum or highest point. Later we will see how this can be useful information.

Decoding Tables and Graphs

Maximum

Minimum

When solving quadratic equations you are solving for the roots (the x values). Pictorially, it is where the parabola crosses the x axis. Algebraically, it is the factored form of the equation or if it cannot be factored, the quadratic formula or a strategy completing the square can be used to solve for the roots or x values.

Let's see how this works. Using the same equation $2x^2 - 4x + 2 = y$, we could set up a table and plug in x values to get y values and then determine where the zeroes are as well as the vertex, but this could take time, whereas factoring the equation could be faster, but notice that both eliminate one-way math.

x	$2x^2 - 4x + 2$	y
-2	$2(-2)^2 - 4(-2) + 2 =$	18
0	$2(0)^2 - 4(0) + 2 =$	2
2	$2(2)^2 - 4(2) + 2 =$	2

First, we notice that the quadractic equation is positive, which tells us that the parabola will be opening upward; therefore, we will have a minimum area or point. Keeping that in mind, we will watch to see where the y values change from descending values to ascending values. Here you can see that there are two y values of 2. This alerts us to investigate further. There must be a value between the 2's that is the vertex. If the y value is equal to 0 then it is. Let's substitute 1 for the x value because it is the next integer between the two 2's:

$2(1)^2 - 4(1) + 2 = y$
$2 - 4 + 2 = y$
$-2 + 2 = y$
$0 = y$

Yes, $(1, 0)$ is the vertex.

We can also tell where the parabola crosses the y axis. The arrow indicates that when $x = 0, y = 2$. So the graph intercepts the y axis at $(0, 2)$. The table above confirms our anaylsis.

Now let's solve this equation algebraically by factoring to find where the graph crosses the x axis.

$2x^2 - 4x + 2 = y$ First, set y to equal 0.
$2x^2 - 4x + 2 = 0$
$\dfrac{2x^2}{2} - \dfrac{4x}{2} + \dfrac{2}{2} = 0$ Now simplify.
$x^2 - 2x + 1 = 0$ Now factor.
$(x - 1)(x - 1) = 0$ Now set each binomial to equal zero.
$(x - 1) = 0 \quad (x - 1) = 0$

Now remove the parentheses and solve for x.

$x - 1 = 0 \quad x - 1 = 0$
$\underline{+1 \quad +1} \quad \underline{+1 \quad +1}$
$x = 1 \quad\quad x = 1$

Notice that the x values are equal. This means that the parabola touches the axis at positive one, therefore, the vertex is $(1, 0)$.

Now let's look at quadratic equations that consist of binomials. How are they graphed? Let's look at the equation $y = (x - 2)^2$. When a quadratic equation is in this form, it has already been factored. Working backward, the original equation is $x^2 - 4x + 4 = y$. Notice that the first term and the last term are perfect squares. In the previous example ($2x^2 - 4x + 2 = y$) we did not simplfy the factored form $(x - 1)(x - 1)$ to be $(x - 1)^2$ because we wanted to make the point that when two binomials are the same, the vertex will be on the x axis. Here the work has been done. All you have to do is interpret the information that is given from the quadratic equation.

When a binomial is squared, as in the example above, just think "opposite direction" than what the sign says inside the parentheses. So the interpretation of this binomial $y = (x -1)^2$ is to move right horizontally one unit starting at the origin. Here are a few examples for interpretation:

The sign is negative so move right 3 units from the origin.
$y = (x - 3)^2$

or

The sign is positive so move left 3 units from the origin.
$y = (x + 3)^2$

Notice the negative in front of the binomial. The procedure still applies.

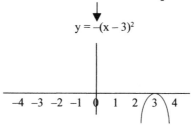
$y = -(x - 3)^2$

If a constant is added to the equation the parabola moves vertically, depending on the sign:

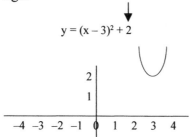
$y = (x - 3)^2 + 2$

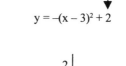
$y = -(x - 3)^2 + 2$

Notice that the binomial is above the x axis; therefore, there is no solution because it does not go through the x axis.

Notice that the binomial goes through the x axis; therefore, the roots or x values are 1.6 and 4.4.

Let's check each equation. The original equation for $y = (x - 3)^2$ can be found by undoing the binomial square $y = (x - 3)(x - 3)$. Use the foil method to rewrite the factors into a trinomial, setting y to equal zero.

$(x - 3)(x - 3) \longrightarrow x^2 - 6x + 9 = y$ Now subsitute the x value into the original
$x = 3$ $x^2 - 6x + 9 = 0$ equation and solve. If both sides are equal,
    $(3)^2 - 6(3) + 9 = 0$ the x values are solutions.
    $9 - 18 + 9 = 0$
    $-9 + 9 = 0$ Here we have confirmed that the double
    $0 = 0$ root is $x = 3$.

The original equation for $y = (x - 3)^2 + 2$ can be found by the same method above, except the constants are added together at the end, making a new trinomial.

$(x - 3)(x - 3) + 2 = y \longrightarrow x^2 - 6x + 9 + 2 = y$ First, set y to equal zero.
    $x^2 - 6x + 11 = 0$ Substitute zero into the y variable, then simplify by collecting like terms.

    $(3)^2 - 6(3) + 11 = 0$ Now subsitute the x values into the
    $9 - 18 + 11 = 0$ original equation and solve. If both
    $-9 + 11 = 0$ sides are equal, the x values are solu-
    $2 \neq 0$ tions. If not, then there is no solution to the equation.

The 2 does not equal 0, so the equation has no solution.

Not all binomials are perfect squares. Let's see how to graph and interpret these types of quadratic equations.

Example: $x^2 - 10x + 21 = y$

x	$x^2 - 10x + 21$	y
−1	$(−1)^2 − 10(−1) + 21 =$	32
→ 0	$(0)^2 − 10(0) + 21 =$	21
1	$(1)^2 − 10(1) + 21 =$	18
2	$(2)^2 − 10(2) + 21 =$	5
3	$(3)^2 − 10(3) + 21 =$	0 ←
4	$(4)^2 − 10(4) + 21 =$	−3
5	$(5)^2 − 10(5) + 21 =$	−4
6	$(6)^2 − 10(6) + 21 =$	−3
7	$(7)^2 − 10(7) + 21 =$	0 ←

First, let's make a table of values to see if we can conclude the vertex and roots, if any, for this equation.

Looking at equation, the y intercept is 21 positive units and x^2 is positive. This means the parabola will be upward; therefore, it will have a minimum point.

The first root is 3. This is confirmed when y equals zero (3, 0). Because a quadratic equation produces two possible solutions, there is one more root that must be identified, which is is (7, 0).

Notice that the section −3, −4 and −3, −4 must be the vertex because between the y values of −3 and −3 the y value decreases to −4 and then starts increasing again.

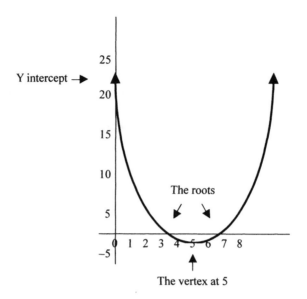

Now let's factor the trinomial to find the roots and then solve for the vertex using the formula $-b/2$.

$x^2 - 10x + 21 = y$
$(x -)(x -) = y$

First, we see that the trinomial has a negative and then a positive sign between the two terms. This tells us that the binomials will both be negative.

Now that we know this, we need to decide what two factors have a sum of $-10x$ but a product of positive 21. The factors of 21 are (1, 3, 7, 21). The only two factors that give a sum of 10 and a product of 21 are 3 and 7.

$(x - 7)(x - 3)$
$x = 7$ and $x = 3$

Check:
$(x - 7)(x - 3)$
$\sqcup$
$-7x$
$+ -3x$
$\overline{-10x}$

Yes, these factors work because the $-10x$ is the middle term in the original trinomial.

Now find the vertex using the formula $-b/2$.

$-(-10)/2 = 5$

This value agrees with the table above (5, −4).

22

Distance, Rate, and Time Problems

INTRODUCTION

Distance, rate, and time problems can be found on many standardized tests. In general, two values are given and the task is to find the third. Approaching these questions in an organized manner is essential. The basic concept for all these problems is that there is a proportion to be satisfied, that is, distance equals the product of rate and time (d = rt).

RELEVANT CONCEPTS FOR ALL TESTS

There is no need to memorize the set of interrelated formulae (example: d/r = t) because all can be derived from the simple d = rt that is applied so universally.

Here is one in the simplest format.

Example: How many miles will a speedboat travel going 80 mph for 2½ hours?

Answer: Outline what is known and what is to be calculated.

d = ?
r = 80 mph
t = 2½ hours

This gives:

d = rt
d = (80)(2½)
d = 200 miles

Slightly harder would be a question in this format:

Example: How long will it take a car averaging 55 mph to travel a distance of 594 miles?

Answer: Let's follow the above format:

$d = 594$

$r = 55$ mph

$t = ?$ (How long indicates time)

Now to plug in what we know:

$d = rt$

$594 = 55t$

$594/55 = t$

$10.8 = t$ or 10.8 hours or 10 hours, 48 minutes

To convert decimal hours into minutes, convert .8 to minutes by changing .8 into a fractional equivalent, then multiply it by 60 because there are 60 minutes in one hour.

$8/10 \cdot 60/1 = 480/10 = 48$ minutes

Sometimes the given data must be rewritten or converted. The next problem requires that.

Example: What is the average speed of a train if it takes three complete days to travel 3,600 miles?

Answer: Again, list what is given.

$d = 3,600$

$r = ?$

$t = 3$ days ⟶ This must be converted to hours because the speed is in *miles per hour*. Twenty-four hours in a day times 3 days = 72 hours.

$t = 72$ hours

Solve for r by dividing both sides by 72:

$3,600/72 = r(72)/72$

$50 = r$ The train averages 50 mph.

Sometimes the problem requires two steps.

Example: A plane flies from New York City to Chicago, a distance of 1,600 miles, at 400 mph. Returning to New York City from Chicago, it flies into a headwind and averages only 320 mph. How many hours total was the plane in the air for the entire trip? (Note: headwind means *against* the wind; tailwind means *with* the wind.)

Answer:

Step 1: New York to Chicago

  d = rt

  d = 1,600

  r = 400

  t = ?

  | New York → Chicago | *Note*: Both ways have equal |
  | Chicago → New York | distances of 1,600 miles. |

  1,600/400 = 400t/400

  4 = t   (4 hours to fly west)

Step 2: Chicago to New York

  d = 1,600

  r = 320

  t = ?

  d = rt

  1,600/320 = 320t/320

  5 = t (5 hours to fly east)

Step 3: Now for the answer to the question:

  Total flying time is 4 hours + 5 hours = 9 hours

Example: Use the table below to answer the following question:

A train moving at a constant speed leaves St. Louis for Sacramento at time = 0. If Sacramento is 1,800 miles from St. Louis, which equation describes the distance from Sacramento at time = t?

  A. d = 60t − 1800
  B. d = 60t
  C. d = 1,800 − 60t
  D. d = 1,800 + 60t

Distance traveled from St. Louis to Sacramento

Time in hours	Distance from St. Louis (miles)
1	60
2	120
3	180
4	240

See that the table is increasing by 60, which is the unit rate for time.

Answer: The basic formula is d = rt, but the question wants to know the constantly decreasing distance from Sacramento, *not* the distance from the starting point. The speed is a constant 60 mph from the chart. So, d = 1,800 – rt or d = 1,800 – 60t. The third answer is correct.

Example: Michelle walks to school every day. She leaves each morning at 7:22 and arrives eighteen minutes later. If she walks at a steady rate of 3.7 mph, what is the distance from her home to the school?

Answer: You must first convert the minutes to hours.

$\quad$ 18 minutes = 18/60 = 9/30 = 3/10 = .3

Now, to answer the question:

$\quad$ D = rt

$\quad$ D = (3.7)(.3) = 1.11 miles

Be careful of questions like this where time is given in minutes. You *must* convert to hours because rate is given in miles per *hour*. One answer will be the result of using unconverted time. It will be a wrong answer.

END OF PRAXIS I, 0014, 0511, AND 0146

Example: Henry drove 100 miles to visit a friend. If he had driven 8 miles per hour faster than he did, he would have arrived in ⁵⁄₆ of the time he actually took. How many minutes did the trip take?

Answer: A chart will help solve this problem.

	d	r	t
Actual trip	100	r	t
Faster trip	100	r + 8	5t/6

Step 1: Because the distance would be the same in both cases, we can set it up as:

$\quad$ rt = rt

$\quad$ rt = (r + 8)(5t/6)

$\quad$ 6rt = (r + 8)(5t)

$\quad$ 6rt = 5rt + 40t

$\quad$ 6rt – 5rt = 40t

$\quad$ rt = 40t

$\quad$ (rt)/t = 40

$\quad$ r = 40 mph

Step 2: Going back to the actual trip:

d = rt

100 = 40t

100/40 = t

5/2 = t

2.5 hours or 150 minutes = time.

There may also be problems involving rates of change that don't actually use the d = rt formula.

Example: Since 1950, when Martin graduated from high school, he has gained 2 pounds each year. In 1980 he was 40% heavier than in 1950. What percent of his 1995 weight was his 1980 weight?

Answer: From 1950 to 1980 (30 years) he gained 30 × 2 lbs. or 60 lbs. That was 40% of his 1950 weight.

30(2) = .4x

60 = .4x

60/.4 = x

150 = x (Martin's weight in 1950)

This gives:

150 lbs. in 1950

210 lbs. in 1980 (+60)

240 lbs. in 1995 (+90)

And thus: 1980/1995 = 210/240 = 21/24 = 7/8 = 87.5%

Example: A pair of climbers decided to hike to the top of Table Rock in South Carolina. They started at 9:00 a.m. and returned at 3:00 p.m. after taking three 5-minute breaks going up and 45 minutes for lunch. Going up they only went 2 mph but returning they could go 3 mph. How long did it take for them to hike to the top, and what was the distance?

Answer: Start with the chart.

	d	r	t
Up Table Rock	same	2	x
Down Table Rock	same	3	5 − x

(Times are x and 5 − x because the excursion lasted 6 hours but there was a 45-minute lunch and three five-minute stops going up, which equals 60 minutes or one hour, and 6 − 1 = 5.)

Equation:

rt = rt

2x = 3(5 − x)

2x = 15 − 3x

2x + 3x = 15

5x = 15

x = 3

Three hours up and two hours down. The distance up (or down) would be 2 mph × 3 hours or 6 miles.

Example: A passenger train from Atlanta to New Orleans starts at 40 mph. Two hours later a freight train leaves from Atlanta for New Orleans at 60 mph. How long will it take before the second train catches up with the first?

Answer:

```
                40 mph + 2 hrs
Atlanta         ─────────────►  New Orleans
                   60 mph
```

	d	r	t
Passenger train	same	40 mph	x
Freight train	same	60 mph	x − 2

Set up as:

rt = rt

40(x) = 60(x − 2)

40x = 60x − 120

40x − 60x = −120

−20x = −120

20x = 120

x = 120/20 = 12/2 = 6

So the passenger train will travel for 6 hours and the freight train for 4 hours. The second train will catch up with the first in 4 hours.

Example: Two planes leave Atlanta at 10:00 a.m., one heading for Rome, Italy at 600 mph, the other for Birmingham, Alabama at 150 mph. At what time will they be 900 miles apart? How far will each have traveled?

Answer:

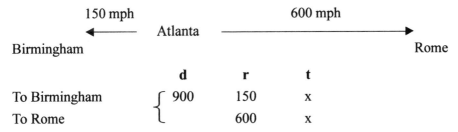

	d	r	t
To Birmingham	⎧ 900	150	x
To Rome	⎩	600	x

Answer: Be careful when clock times are used!

Set up as:

rt + rt = 900

150x + 600x = 900

750x = 900

x = 900/750 = 90/75 = 6/5 = 1^1/₅ hours = 1 hour, 12 minutes.

They will be 900 miles apart at 10 + 1 hour and 12 minutes or 11:12 a.m. The slow plane will have gone (150)6/5 = 180 miles and the fast plane will have gone (600)6/5 = 6(120) = 720 miles.

Example: Ralph and Roger took a motorcycle trip. After lunch, Ralph started to town. Ten minutes later, Roger followed. If Ralph was going 30 mph and Roger 35 mph, how long did it take before Roger caught up with Ralph?

Answer:

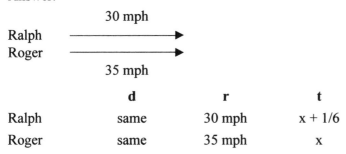

	d	r	t
Ralph	same	30 mph	x + 1/6
Roger	same	35 mph	x

Remember, 10 minutes must be stated in terms of hours because the speed is in miles per hour, so 10/60 = 1/6 of an hour.

Set up as:

rt = rt

30(x + 1/6) = 35x

30x + 5 = 35x

30x − 35x = −5

−5x = −5

$5x = 5$

$x = 5/5 = 1$ hour

Example: James traveled 120 miles driving at a steady speed. Had he increased his speed by 15 mph he could have completed the trip in $2/5$ of an hour less. How fast did he actually go?

Answer:

	d	r	t
Actual trip	120 miles	x	120/x
Faster trip	120 miles	x + 15	120/(x + 15)

Set up as:

$120/x = (120/(x+15)) + 2/5$ (LCD = $5x(x+15)$)

$5(120)(x + 15) = 5x(120) + 2x(x + 15)$

$600x + 9,000 = 600x + 2x^2 + 30x$

$-2x^2 - 30x + 9,000 = 0$

$x^2 + 15x - 4,500 = 0$

$(x + 75)(x - 60) = 0$

$x + 75 = 0 \quad\quad x - 60 = 0$

$x = -75 \quad\quad x = 60$

The negative answer is not possible, so the answer is 60 mph.

23

Work Problems

INTRODUCTION

Work problems are troublesome for many people. A little practice will show you how to cope with these questions.

RELEVANT CONCEPTS FOR ALL TESTS

Example: Eric can plow a field alone in four hours. It takes Sam five hours to plow the same field alone. If they work together and each has a plow, how long will it take to plow the field?

Answer: Set up the equation:

$1/t = 1/4 + 1/5$

$1/t = 5/20 + 4/20$

$1/t = 9/20$

$20 = 9t$

$20/9 = t$

$2\,2/9 = t$

So it will take $2\,2/9$ hours for Sam and Eric to plow the field together.

An alternative approach uses a slightly different, but simpler formula and arrives at the same answer.

$1 = 1t/4 + 1t/5$

$1 = 5t/20 + 4t/20$

$1 = 9t/20$

$^{20}/_9 = t$

$2^2/_9 = t$ (Yields the same answer)

Example: Jerry can mow John's lawn in three hours. Bob can mow it in six hours. If John hires Jerry and Bob to work together using two mowers, how fast can they mow the lawn?

Answer: Set up the equation:

$^1/_t = ^1/_3 + ^1/_6$

$^1/_t = ^2/_6 + ^1/_6$

$^1/_t = ^3/_6$

$^1/_t = ^1/_2$

$t = 2$ hours

Or, using the second approach,

$^{1t}/_3 + ^{1t}/_6 = 1$

$^{2t}/_6 + ^{1t}/_6 = 1$

$^{3t}/_6 = 1$

$^t/_2 = 1$

$t = 2$ Which agrees with the answer above.

Example: Jack can paint a house in eight hours. Sean can paint the house in six hours. Oscar can also paint the house in six hours. How long will it take if they all work together?

Answer: Set up the equation:

$^1/_t = ^1/_8 + ^1/_6 + ^1/_6$

$^1/_t = ^3/_{24} + ^4/_{24} + ^4/_{24}$

$^1/_t = ^{11}/_{24}$

$24 = 11t$

$^{24}/_{11} = t$

$t = 2^2/_{11}$ hours

Or, using the other format,

$1 = ^{1t}/_8 + ^{1t}/_6 + ^{1t}/_6$

$1 = ^{3t}/_{24} + ^{4t}/_{24} + ^{4t}/_{24}$

$1 = ^{11t}/_{24}$

$^{24}/_{11} = t$

$t = 2^2/_{11}$ hours

Here is a different twist on a work problem.

Example: A tank is being filled at a rate of ten gallons per hour. However, a hole in the tank allows water to run out at a rate of two gallons per hour. How long will it take to fill an empty fifty-gallon tank?

Answer: Each hour the tank gains 10 − 2 or 8 gallons. So to fill the tank will take 50/8 hours, which is $6^{1}/_{4}$ hours.

Example: Sue, Betty, and Moira decide to type Judy's term paper. Sue can type three chapters per hour, Betty can type five chapters per hour, and Moira can type six chapters per hour. If Judy's term paper is thirty chapters long, how long will it take for all of them working together to type the entire paper?

Answer: The three girls can type 3 + 5 + 6 or 14 chapters per hour. The job will take 30/14 hours or $2^{2}/_{14}$ hours, which reduces to $2^{1}/_{7}$ hours.

Example: Micki and Vicki can restock a supermarket aisle in one hour working together. Micki can restock the aisle in an hour and a half working alone. Vicki can restock the aisle in 2 hours working alone. If they work together for two hours and then work separately for another two hours, how many aisles can they restock?

Answer: In the first two hours they can complete 2 aisles (1/hour). Micki can complete 2/1.5 aisles in two hours (1.33 aisles). (You ask yourself, "How many 1.5 *units* are there in 2 hours?") Vicki can complete another aisle in two hours. This gives a total of 2 + 1.33 + 1 = 4.33 (or $4^{1}/_{3}$) aisles in 4 hours.

END OF PRAXIS I, 0014, 0511, AND 0146

Now for a sample of problems that are a bit more complex.

Example: Two printing presses working together can complete a job in 2.5 hours. Working alone, press A can do the job in 10 hours. How many hours will press B take to do the job by itself?

Answer: Set up the equation:

   1 = 2.5/10 + 2.5/x

Clear the fractions by multiplying both sides by 10x:

   10x = 2.5x + 25
   10x − 2.5x = 25
   7.5x = 25
   x = 25/7.5
   x = 3.3

It will take press B only 3.3 hours to complete the job.

Example: Mr. Dodgen hires Jim and John to trim his peach trees. Jim can do the job in 50 hours but John can do it in only 40. How many hours will it take them working together?

Answer: Set up the equation:

$1/50 + 1/40 = 1/x$

Multiply by 200x to clear the fractions:

$4x + 5x = 200$

$9x = 200$

$x = 200/9 = 22\,{}^2\!/_9$ hours

Example: Two students, George and Tim, agree to paint a barn. George can paint it alone in 5 days and Tim can paint it alone in 8 days. They start painting together but after two days Tim quits, so George has to finish alone. How long does it take for the barn to be painted? How long does George have to paint after Tim quits?

Answer diagram:

	Days	**Done in One Day**
George	5	$1/5$
Tim	8	$1/8$

Formula:

George in 2 days + Tim in 2 days + George to finish

$2/5 + 2/8 + x/5 = 1$

To clear the fractions, multiply by 40:

$8(2) + 5(2) + 8(x) = 40$

$16 + 10 + 8x = 40$

$8x = 40 - 16 - 10$

$8x = 14$

$x = 14/8 = 7/4 = 1\,{}^3\!/_4$

George has to work alone for $1\,{}^3\!/_4$ days, so the complete job takes $3\,{}^3\!/_4$ days.

24

Elementary Algebra

INTRODUCTION

For many students, the word "algebra" evokes images of confusion and difficulty. For others, the word brings up images of endlessly interesting puzzles to be solved by the application of rules and logic. Now also imagine which group does better on standardized tests involving algebra?

RELEVANT CONCEPTS FOR ALL TESTS

Most often, algebra involves using and acting on letters as if they were numbers. There is one fundamental rule that governs most algebraic procedures: "Whatever you do to one side of an equation, do the same to the other."

So, if you add $25 to one side, you must add $25 to the other side in order to keep both sides equal. The same goes for the other operations (subtraction, multiplication, and division). Both sides must remain equal. This is a very important idea in finding a value that will keep both sides of an equation equal at all times. But what if the value is not so easy to determine? How do we find that value? We can write an equation using letters to stand for the missing value that will maintain equality. These letters are called *variables*. Once this is done, we can use a process that will undo the equation. This process, called "isolating the variable," is analogous to unraveling a mystery. When the variable is isolated, the answer will be found. This result can be checked by substituting the value back into the original equation to verify equality. Before showing how to isolate the variable in an equation, let's review the order of operations and see how that relates to this process.

ORDER OF OPERATIONS

Strongly related to the "nature of numbers" is the way we must deal with numbers, which is embodied in the *order of operations*. Certain conventions have been adopted to assure that the same result will always be obtained in calculations. The order of operations is applied sequentially from 1 to 6, moving from left to right in an *expression*. Students often use the mnemonic device "Please Excuse My Dear Aunt Sally" to recall the order (see Chapter 2).

1. Parentheses
2. Exponents (Powers) and roots
3. Multiplication/Division (whichever comes first, left to right)
4. Addition/Subtraction (whichever comes first, left to right)

Example:

Expression = Expression
$(3 + x)^2 - (5x + 3x) =$?

Answer:
$(3 + x)^2 - (5x + 3x)$
$(3 + x)(3 + x) - (8x)$
$9 + 6x + x^2 - 8x$
$9 - 2x + x^2$
$x^2 - 2x + 9$ (rewritten in descending degrees)

Here we are simplifying one side of an equation (i.e., an expression) using the order of operations. If we had two sides of the equation, meaning two expressions equaling each other in some form, we would simplify both sides of the equation using the order of operations before working across the equals sign.

Now, how is the order of operations related to the process of isolating the variable to maintain equality? Remember, in order to find the value we must undo, or unravel, the equation. More precisely, to solve an algebraic problem, your procedure is to reverse the normal order of operations because we are no longer simplifying an expression on one side of an equation, we are now performing operations that require working across the equals sign. The reversed order is:

1. Negate subtraction or addition
2. Negate division or multiplication

3. Negate exponents or radicals

4. Clear parentheses

So the tactic is to use the order of operations in reverse. How is this done?

The simplest equations are called *one-step equations*. These equations will have only one variable on one side of the equation. We start on the side of an equation with the expression that *includes* the variable. Is it being added, subtracted, multiplied, or divided?

For example, subtraction reverses addition (and vice versa) and division negates multiplication. This is all done on one side of the equation, while on the other side of the equation the same value is either being added, subtracted, multiplied, or divided to maintain equality. It is always possible to use letters in these operations as well as numbers.

Example: What is the value of x in x + 3 = 15?

Answer:

$$x + 3 = 15$$
$$x + 3 - 3 = 15 - 3$$
$$x = 12$$

Check:

$$x + 3 = 15$$
$$(12) + 3 = 15$$
$$15 = 15$$

Notice that the 3 on the left side of the equals sign is negated by using the inverse operation, subtraction, leaving only the x, while the 3 on the right side is subtracted from 15, resulting in x equaling 12. Remember, what is done to one side of the equation must be done exactly the same way on the other side to maintain equality.

Because the value of x is 12, substitute 12 for the x variable to see if the mathematical statement is true.

The *property of equality* holds because both sides of the equation equal the same value, 15.

Example: What is the value of x in x − 5 = 20?

Answer:

$$x - 5 = 20$$
$$x - 5 + 5 = 20 + 5$$
$$x = 25$$

Check:

$$x - 5 = 20$$
$$(25) - 5 = 20$$
$$20 = 20$$

Notice that the 5 on the left side of the equals sign is negated by using the inverse operation of addition, leaving only the x, while the 5 on the right side is added to 20, resulting in x equaling 25. Again, what is done to one side of the equation must be done exactly the same way on the other side to maintain equality.

Because x equals 25, substitute 25 for the variable x to see if the mathematical statement is true.

The property of equality holds because both sides of the equation equal the same value, 20.

Example: What is the value of y in $3y = 33$? (Note: 3y means 3 times y.)

Answer:
$$3y = 33$$
$$3y/3 = 33/3$$
$$y = 11$$

Check:
$$3y = 33$$
$$3(11) = 33$$
$$33 = 33$$

Notice that the three on the left side of the equals sign is negated by using division (the inverse of multiplication), leaving only the y, while the 3 on the right side is divided into the 33, resulting in y equaling 11.

Because y equals 11, substitute 11 for the variable y to see if the mathematical statement is true.

The property of equality holds because both sides of the equation equal the same value, 33.

Example: Solve for z: $z/10 = 50$

Answer:
$$(10)z/10 = (10)50$$
$$z = 500$$

Check:
$$z/10 = 50$$
$$500/10 = 50$$
$$50 = 50$$

Notice that the 10 on the left side of the equals sign is negated by using the inverse operation of multiplication, leaving only the z, while the 50 on the right side is multiplied by the 10, resulting in z equaling 500.

Because z equals 500, substitute 500 for the variable z to see if the mathematical statement is true.

The property of equality holds because both sides of the equation equal the same value, 50.

The example below is a more elaborate one-step equation.

Example: What is the value of x in the equation $7 - 8 + 5 + x + 6 - 8 = 40$?

Answer:
1. $7 - 8 + 5 + x + 6 - 8 = 40$
2. $-1 + 5 + 6 - 8 + x = 40$
3. $4 + 6 - 8 + x = 40$
4. $10 - 8 + x = 40$
5. $2 + x = 40$

1–5. Because both adding and subtracting are on the same side of the variable, simplify the expression using *the normal* order of operations to equal one number before working across the equals sign. *Notice how each expression is rewritten using the order of operations from left to right until the left-hand side is simplified to one number and a variable.* Now the variable is isolated by performing the order of operations in reverse.

Elementary Algebra

6. $2 + x = 40$
   $\underline{-2 \qquad -2}$
   $\qquad x = 38$

Check:
$7 - 8 + 5 + (38) + 6 - 8 = 40$
$40 = 40$

6. Now isolate the variable by negating the 2 on the left side. Because the 2 is subtracted from the left side, 2 must be subtracted from the right side, giving the value of x.

   Now check to see if the property of equality holds by substituting the value of x into the equation.

The equation below is considered a multi-step equation:

Example: What is the value of x in the equation $3x - 14 = x + 2$?

Answer:
1. $3x - 14 = x + 2$
   $\underline{-x \qquad\quad = -x}$
   $\;2x - 14 = 2$
2. $\underline{\qquad +14 = +14}$
   $\;2x \qquad = 16$
3. $2x/2 = 16/2$
   $x = 8$

Check:
$3(8) - 14 = (8) + 2$
$24 - 14 = 10$
$10 = 10$

1. Notice that there is an x on both sides of the equation. Because the variable is the same, we can isolate the variable by subtracting the x from both sides of the equation to preserve equality.

2. The x is still not isolated; therefore, we need to add 14 to both sides of the equation to continue preserving equality. Now the equation is $2x = 16$.

3. The unknown is still not isolated; therefore, this time we need to divide by 2 on both sides of the equation to preserve equality. Finally, x equals 8.

   Now check to see if the value x is a true solution.

Example: Solve for x: $^{40}/_x = 5$

Answer:
1. $^{40}/_x = 5$
   $(^1/_{40})^{40}/_x = 5(^1/_{40})$
2. $^1/_x = ^5/_{40}$
   $(1)40 = (5)x$
   $40 = 5x$
3. $^{40}/_5 = ^{5x}/_5$
4. $5(5x) = 40(5)$
   $25x = 200$
   $^x/_{25} = ^{200}/_{25}$
   $x = 8$

Don't let this equation confuse you just because the denominator is a variable. It is easier than the other equations because you can use proportionality to solve it (see Chapter 13, "Ratios and Proportions").

1. Multiply each side of the equation by $^1/_{40}$ and simplify.

2. Now cross-multiply. You can do this because the fractions are between an equals sign. The equation is now $40 = 5x$.

3–4. Now isolate the variable by dividing both sides by 5 and then simplify. The solution is $x = 8$.

Check:

$^{40}/_8 = 5$

$5 = 5$

Now check the answer.

Example: Solve for x: $^{7x}/_8 = 21$

Answer:

1. $^{7x}/_8 = 21$

   $(8)^{7x}/_8 = (8)21$

2. $7x = 168$

   $^{7x}/_7 = ^{168}/_7$

3. $x = 24$

1. Multiply both sides of the equation by 8.
2. Now divide by 7 to isolate x.
3. x equals 24.

Now check the answer.

Check:

$^{7(24)}/_8 = 21$

$^{168}/_8 = 21$

$21 = 21$

Example: Solve for x: $20 = ^x/_3 - 10$

Answer:

1. $3(20) = 3(^x/_3) - 10(3)$

   $60 = x - 30$

   $60 + 30 = x - 30 + 30$

   $90 = x$

2. $20 = ^x/_3 - 10$

   $\underline{+10 \qquad +10}$

   $30 = ^x/_3$

   $(3)30 = ^{x(3)}/_3$

   $90 = x$

There are several ways to isolate the variable in this equation.

1. Multiply through by 3 to give the equation $60 = x - 30$, then isolate the variable by adding 30 to both sides to give the solution $90 = x$.

2. Another way would be to add 10 to both sides of the equation first, then multiply both sides by 3 to give the same solution: $90 = x$.

Finally, check the answer.

Check:

$20 = ^x/_3 - 10$

$20 = ^{90}/_3 - 10$

$20 = 30 - 10$

$20 = 20$

Elementary Algebra

The following is an example of a word problem.

Example: How many girls are there in a class of 30 pupils if the number of boys is 10 less than the number of girls?

Answer:

$g + b = 30;\ b = g - 10$

$g + b = 30$

$g + (g - 10) = 30$

$2g - 10 = 30$

$\underline{ + 10\ +10}$

$2g = 40$

$2g/2 = 40/2$

$g = 20$

Number of girls is 20
(The substitution method)

What do we know? We know that girls + boys = 30, which can be expressed as $g + b = 30$. We also know that boys = girls – 10, which can be expressed as $b = g - 10$.

Now substitute the expression for b in the $g + b = 30$ equation. Then collect like terms. There are two g's, so the rewritten equation is $2g - 10 = 30$.

Now add 10 to both sides of the equation, making the equation $2g = 40$.

Now isolate the variable by dividing both sides of the equation by 2 to find the value of g, which is 20. Because there are a total of 30 pupils in the class, 20 of the pupils are girls. Although the question did not ask for the number of boys, it is easy to determine by subtracting the number of girls from the total number of students.

This gives 10 boys.

Here is another approach to solving this word problem.

girls = girls
$b + 10$ $30 - b$
(The Proportionality Method)

Collect like terms. There are two b's, so the rewritten equation is $2b + 10 = 30$.

Now subtract 10 from both sides of the equation, making the equation $2b = 20$, $b = 10$.

What do we know? We know that girls = 30 pupils – boys. We also know that girls = boys + 10 (instead of boys – 10). Now we can write an equation that will be in terms of boys. This means that each expression is written to solve for boys (b).

If the word problem says that the boys are 10 less than the number of girls, then it makes sense that the total of girls would be boys + 10. The word problem also says that there are 30 pupils in the class. Then it would also be true to say that girls would equal 30 minus the boys $(30 - b)$.

Now isolate the variable by dividing by 2 to find the value of b, which is 10. Because there are 30 pupils in the class, and 10 of the pupils are boys, there must be 20 girls in the class.

Notice that both expressions equal the number of girls. This allows us to set them equal to each other. Now all we have to do is to collect like terms and solve for b. This will give us the number of boys instead of girls. Now just subtract the number of boys from 30 to find the number of girls. Of course, there are other ways as well. Remember, math is not done only one way.

Example: Charles has $37.00 in his bank account for Christmas and hopes to increase it to $100 by making equal deposits each week. How much should he deposit if he has 14 weeks?

Answer:

$(100 - 37)/14 = x$

$63/14 = x$

$4.5 = x$

$\$4.50 = x$ Charles should deposit $4.50 each week for 14 weeks.

Check:

$(100 - 37)/14 = 4.50$

$63/14 = 4.50$

$4.50 = 4.50$

Example: If $p \div 5 = q$, what is $p \div 10$?

Answer:

$p \div 5 = q$ $p \div 10 = x$

1. $p \div 5 = q$

   $p/5 = q$

2. $(5)p/5 = (5)q$

3. $p = 5q$

4. $5q \div 10 = x$

   $5q/10 = x$

   $q/2 = x$

   or $q \div 2$ (rewritten)

This is a perfect example to describe several mathematical strategies.

First approach: We could look at both expressions and ask, "what do these two expressions have in common?" They both have the variable p in them. We could use that information to find the value of p, then substitute that value for both p's to find how they are related. In order to use p as a tool, we will rewrite the $p \div 5 = q$ in terms of p.

1. Rewrite the equation so that it will be easier to manipulate by placing the p as the numerator.

2. Now isolate the p by multiplying both sides by 5.

Elementary Algebra

1. $5q = p$
2. $5q = p$ and $p \div 10 = ?$
3. $5q/10 = p/10$
   $1q/2 = p/10$
   $q \div 2 = p \div 10$

$p \div 5 \qquad p \div 10$
$(10) \div 5 \qquad (10) \div 10$
$2 \qquad\qquad 1$

3. Now simplify, leaving the equation $p = 5q$. Now the equation is in terms of p.
4. Now take the expression $p = 5q$ and substitute 5q into the other expression, $p \div 10 = ?$ The easiest way is to rewrite this expression as a fraction, then simplify. It equals $q \div 2$.

Second approach: Recognize that 5 and 10 are both related, and then use this information along with the property of equality to find what $p \div 10$ equals.

1. Rewrite $p \div 5 = q$ in terms of p by following the steps above. The result is $5q = p$.
2. 5q equals p and $p \div 10$ equals some number.
3. Now think about how 10 can be used to simplify the first equation, $5q = p$, using the property of equality. We can see that 5 is a factor of 10; therefore, if I take the 10 and divide into both sides of the $5q = p$ equation, it will simplify the left side to equal $q/2 = p/10$, which can be rewritten as $q \div 2 = p \div 10$.

Third approach: simply substitute numbers for the p's to get a ratio between the two expressions. The only catch is, we have to make sure we are answering the question that is being asked. Here the question is asking to solve for $p \div 5$. For example, we could substitute 10 for p in each of the equations because 5 is a factor of 10 (we could have used equivalent numbers like 20, 30, etc.). Remember, we are only looking for an easy way to find a solution).

Now simplify. This makes a ratio of 2:1 so that moving from p/5 to p/10, the quotient is halved (in the ratio of 2:1). Now the solution can be written as q/2 or $q \div 2$. Both express the fact that the quotient is half as much because q/2 is $1/2(q)$.

END OF PRAXIS I, 0014, 0511, AND 0146

Example: What is the value of ab if $a^2 + b^2 = 4$ and $(a - b)^2 = 2$?

Answer:

$(a - b)^2 = 2 \quad\quad a^2 + b^2 = 4$

$\downarrow \quad\quad\quad\quad\quad \downarrow$

$a^2 - 2ab + b^2 = 2 \quad a^2 + b^2 = 4$

$\quad\quad a^2 \quad\quad + b^2 = 4$

$\underline{- (a^2 - 2ab + b^2 = 2)}$

$\quad\quad\quad 2ab \quad\quad = 2$

$2ab/2 = 2/2$

$ab = 1$

Two variables equal 1? This is different. What could the a and b equal in order for the product to equal 1? Could they be reciprocals of each other? Yes.

Can you reason through this example using what you know about algebraic manipulation?

Can we rewrite $(a - b)^2$ as $a^2 + b^2$? No, but this is an important idea to see if you know how these two statements are manipulated.

For example: If $a = 3$ and $b = 4$.

$(3 - 4)^2 \neq 3^2 + 4^2$

$\downarrow \quad\quad\quad \downarrow$

$-1^2 \quad \neq 9 + 16$

$\downarrow \quad\quad\quad \downarrow$

$1 \quad\quad \neq 25$

$(a - b)^2$ means there are two binomials that are the same and they must be multiplied. The FOIL method may be used to give a trinomial.

> The FOIL method is for calculating the product of two binomials. F is the product of the first terms, O is the outside terms, I is the inside terms, and L is the last terms.
>
> For example: $(a - b)(a - b) =$
>
> $F = a^2$
>
> $O = -ab$
>
> $I = -ab$
>
> $L = b^2$
>
> This simplifies to $a^2 - 2ab + b^2$.

Because they have common factors, I can subtract the two equations, making sure that I keep all like terms together. Now divide by 2 on both sides to find what ab equals.

Another approach is to substitute one equation for the other.

$(a - b)^2 = 2 \quad\quad a^2 + b^2 = 4$

$\downarrow \quad\quad\quad\quad\quad \downarrow$

$a^2 - 2ab + b^2 = 2 \quad a^2 + b^2 = 4$

This manipulation makes you think outside of the box.

$a^2 + b^2 - 2ab = 2$

$\underbrace{}$

$4 - 2ab = 2$

$\underline{-4 -4}$

$ -2ab = -2$

$-2ab/2 = -2/2$

$ab = 1$

Notice that the equation $a^2 - 2ab + b^2 = 2$ can be rewritten as $a^2 + b^2 - 2ab = 2$. This can be done using the commutative property. It is given that $a^2 + b^2$ equals 4 from the second equation; therefore, I can substitute 4 for $a^2 + b^2$ (see the manipulation on the left under the bracket). Now the equation is $4 - 2ab = 2$. Solve for ab by subtracting 4 from both sides, which gives $-2ab = -2$. Then divide both sides by -2. The solution is $ab = 1$.

Example: If r and s are two non-zero numbers and if $78(r + s) = (78 + r)s$, which of the following MUST be true?

r = 78   s = 78   r + s = 78   r < 1   s < 78

Answer:

$78(r + s) = (78 + r)s$

$78r + 78s = 78s + rs$

$78r = rs$

$78\cancel{r}/\cancel{r} = \cancel{r}s/\cancel{r}$

$78 = s$ (This is the second choice above.)

1. Use the distributive property on both sides of the equation to simplify.
2. Collect terms by using the property of equality.
3. Isolate s by dividing both sides by r.
4. Simplify all r's to obtain $78 = s$.

Example: If $f(x) = x^2 - 2^x$, what is the value of $f(4)$?

Answer:

$f(x) = x^2 - 2^x$

$f(4) = 4^2 - 2^4$

$f(4) = 4 \cdot 4 - (2 \cdot 2 \cdot 2 \cdot 2)$

$f(4) = 16 - 16$

$f(4) = 0$

Notice that $f(x)$ really is another notation for $y = x^2 - 2^x$. Here, if 4 is substituted for x, what will be the y value? Zero.

Solving Equations Containing Absolute Value

Examples:

1. $|-5| =$
2. $|5| =$
3. $|6 - 9| =$
4. $5 - |-7| =$

Notice that each equation is asking for the absolute value. It is *not* asking to solve for $|x|$. Thus, the absolute value of a number is always positive except when the number is zero.

$|0| = 0$ value always zero

$\left.\begin{array}{l}|x| > 0 \\ |-x| > 0\end{array}\right\}$ values are always positive

Answer:
1. $|-5| = 5$
2. $|5| = 5$
3. $|6 - 9| = |-3| = 3$
4. $5 - |-7| = 5 - 7 = -2$

If there is one term inside the absolute value bars, then the value is always positive. See examples 1 and 2.

If there are at least two terms inside the absolute value bars, simplify before getting the absolute value. See example 3.

If there is a negative sign outside the absolute value bars, keep the sign. See example 4.

Example: $3|x| + 5 = 23$ (*Note*: $|x|$ denotes the absolute value of x.)

Answer:
$3|x| + 5 = 23$
$ -5 \ -5$
$\overline{3|x| = 18}$
$\cancel{3}|x|/\cancel{3} = 18/3$
$|x| = 6$
$x = 6$ and $x = -6$

Although absolute value gives a positive solution, the integer between the absolute value bars may be negative or positive because absolute value refers to the distance from zero. Therefore, an absolute value equation will have two solutions. This means negative and positive values for x should be considered.

Example: Solve this absolute value compound equation: $|4x + 2| = 10$.

$\|4x + 2\| = 10$	or	$\|4x + 2\| = -10$
$4x + 2 = 10$		$4x + 2 = -10$
$-2 \ -2$		$-2 \ -2$
$4x = 8$		$4x = -12$
$4x/4 = 8/4$	or	$4x/4 = -12/4$
$x = 2$	or	$x = -3$

1. Because absolute value refers to the distance from zero, 10 represents the number of units away from zero; however, it is not a simple absolute value of x. Rather, it is the absolute value of a *binomial*, which means that the equation will not have the same positive and negative x values to consider.

2. Remember that absolute value has two parts, so two equations have to be written. One equation will equal positive 10 units and the other will be negative

Check:
|4(2) + 2| = 10 |4(−3) + 2| = 10
|10| = 10 |−10| = 10
10 = 10   10 = 10

10 units. The equation can be translated to show that 4x + 2 is 10 units from zero on a number line.

3. Solve both equations for x by dividing both sides by 4 in each equation. The solution set is (2, −3). This set represents x values of −3 and 2.

4. Check by substituting the x values in the original equation.

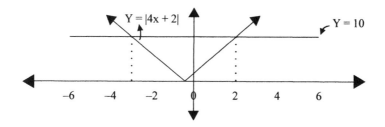

How would you solve this absolute value compound equation?

Example: $5|x − 1| − 2 = 13$

1. $5|x − 1| − 2 = 13$
   $ +2 +2$
   $\overline{5|x − 1| = 15}$

2. $5|x − 1|/5 = 15/5$

3. $|x − 1| = 3$ or $|x − 1| = −3$
   $x − 1 = 3$ $x − 1 = −3$
   $ + 1 + 1$ $ + 1 + 1$
   $\overline{x = 4}$ $\overline{x = −2}$

The answer set is (4, − 2).

This example has more than the absolute value of a binomial. The equation has to be rewritten so that the absolute value is isolated on one side before solving the compound equation.

1. Isolate the absolute value by using the order of operations in reverse. So add 2 to both sides to maintain equality.

2. Now isolate the |x − 1| by dividing both sides of the equation by 5.

3. Now set up a compound equation without the absolute value bars, with one equation equaling positive 3 and the other equaling negative 3. Then solve each equation for x.

4. Now check to see if both x values are valid solutions.

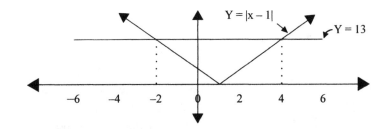

4. Check: $5|4 - 1| - 2 = 13$ $\quad\quad 5|-2 - 1| - 2 = 13$
 $5|3| - 2 = 13$ $\quad\quad\quad\quad 5|-3| - 2 = 13$
 $5(3) - 2 = 13$ $\quad\quad\quad\quad 5(3) - 2 = 13$
 $15 - 2 = 13$ $\quad\quad\quad\quad\, 15 - 2 = 13$
 $13 = 13$ $\quad\quad\quad\quad\quad\quad 13 = 13$

Both solutions are correct.

Example: Is this absolute value equation $|6x - 1| = -3$ true or false?

$|6x - 1| = -3$ or $|6x - 1| = -3$

$\begin{array}{r}6x - 1 = -3 \\ +1 \quad +1\end{array}$ or $\begin{array}{r}6x - 1 = 3 \\ +1 \quad +1\end{array}$

$6x/6 = -2/6 \quad\quad\quad\quad 6x/6 = 4/6$

$x = -1/3 \quad\quad\quad\quad\quad x = 2/3$

Check:

$|6(-1/3) - 1| = -3$ or $|6(2/3) - 1| = -3$

$|-2 - 1| = -3 \quad\quad\quad\quad |4 - 1| = -3$

$|-3| = -3 \quad\quad\quad\quad\quad\, |3| = -3$

$3 \neq -3 \quad\quad\quad\quad\quad\quad 3 \neq -3$

No solution

It is false. This equation really did not need to be solved because the original equation is equal to a *negative* value, which cannot happen. Remember that absolute value means that the solution is always positive or zero (see the original equations above). Each has a positive solution. However, to prove that this is a false statement, let's find the set of (– and +) x values by solving of both parts. Now substitute the x quantities into the original equation. What happens? The original equation is equal to a negative 3, whereas the absolute value equals a positive 3, which means there *is no* solution for the absolute value equation that equals a negative value.

Make sure to pay attention to what is being asked.

Rule: The absolute value of an expression can never be negative. Notation is written like this: $|a| \geq 0$ for all real numbers a.

Example: $|5x - 3| = 0$ (This mathematical statement is saying that the absolute value of 5x − 3 equals zero.) When is this statement true?

$|5x - 3| = 0$

$5x - 3 = 0$

$ + 3 +3$

$\overline{5x = 3}$

$5x/5 = 3/5$

$x = {}^3/_5$

What value for x will make this expression equal zero? Three-fifths ($^3/_5$) is the only solution that will satisfy this equation. Any other solution will make this equation false.

Check:

$|5x - 3| = 0$

$|5(3/5) - 3| = 0$

$|3 - 3| = 0$

$|0| = 0$

$0 = 0$

Rule: The absolute value of an expression equals 0 only when the expression is equal to 0.

Solving an Equation with Two Absolute Values

Example: $|z + 12| = |3z - 9|$

$|z + 12| = |3z - 9|$ or $|z + 12| = -|3z - 9|$

$z + 12 = 3z - 9$ $z + 12 = -3z + 9$

$-z -z$ $-z -z$

$\overline{ 12 = 2z - 9}$ $\overline{ 12 = -4z + 9}$

$ +9 +9$ $ -9 -9$

$\overline{ 21 = 2z}$ $\overline{ 3 = -4z}$

$21/2 = 2z/2$ $3/-4 = -4z/-4$

$10.5 = z$ $-{}^3/_4 = z$

Notice that the second equation has a negative sign outside the binomial. Why? Because absolute value has two parts, in this case a negative z and a positive z, and both have to be considered in order to find the solution set for z (10.5, $-{}^3/_4$). This solution set is not the same as previous examples. Instead, the z values here are where the two graphs intersect. This is because there are two absolute value expressions that are equal to each other at certain points. In essence, the mathematical statement asks whether the two graphs at some point have the same x value.

Check:

$|10.5 + 12| = |3(10.5) - 9|$ or $|-3/4 + 12| = |3(-3/4) - 9|$
$\quad |22.5| = |31.5 - 9| \qquad\qquad |11.25| = |-11.25|$
$\quad 22.5 = 22.5 \qquad\qquad\qquad 11.25 = 11.25$

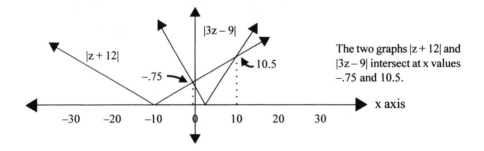

The two graphs $|z + 12|$ and $|3z - 9|$ intersect at x values $-.75$ and 10.5.

Sometimes a problem will involve solving two equations containing two unknowns. These equations are often called systems of linear equations. Three things must be true before this approach may be used:

1. The two equations must contain the same two variables.

Notice here that there will be a solution set y = or x = some number. Think of intersecting lines.

2. They *cannot* have the same slope ($3x + 6$ and $3x + 5$). The sum of these of equations will have no solution like this one ($0 \ne 1$) because the two lines are parallel and will never intersect. The solution indicates the system is inconsistent.

Notice here that there will be no solution. Think of parallel lines.

3. The equations cannot be identical ($2x - y = 3$ and $6x - 3y = 9$). The sum of equivalent equations will be $0 = 0$, which is a true statement, but they are considered dependent equations because both equations lie on the same line.

Notice here that the solution will always be $0 = 0$, which is a true statement. Think of equivalent lines; therefore, there is an infinite number of solutions.

Elementary Algebra

The purpose for this type of manipulation is to find the point where the two equations are equal or graphically where the two lines intersect. There are two approaches: the *elimination method* and the *substitution method*.

Example: $x + y = 4$
$2x - y = -1$

Answer using the elimination method, where the two equations are added together to eliminate one variable.

$x + y = 4$
$+ (2x - y = -1)$
$\overline{3x = 3}$

$3x/3 = 3/3$

$x = 1$

$x + y = 4$
$(1) + y = 4$
$1 + y = 4$
$\underline{-1 -1}$
$y = 3$

The elimination method is useful when both equations have the same variable with opposite signs (see the example on the left). Notice that the y variables have opposite signs. This makes adding the two equations simple because the y's negate themselves, leaving the rest of the equation easy to solve using the one-step process.

Now substitute 1 into the x value of the first equation to solve for the y value.

The solution is $x = 1, y = 3$.

The x and y values are an ordered pair (1, 3), which indicates the "point" where both equations are equal and where their lines intersect.

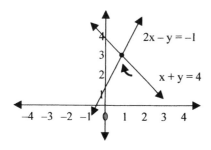

The two lines intersect at one point (1, 3).

Using the same equations: $x + y = 4$
$2x - y = -1$

Answer using the substitution method:

1. $x + y = 4$ First equation
    $\underline{-x -x}$
    $y = 4 - x$

    The substitution method is used when one of the variables x or y has a coefficient of 1. In this example, y will work easily.

2. $2x - y = -1$ Second equation
   $\downarrow$
   $2x - (4 - x) = -1$
3. $2x - 4 + x = -1$
   $3x - 4 = -1$
   $3x - 4 + 4 = -1 + 4$
   $3x = 3$
   $x = 1$
4. $x + y = 4$
   $(1) + y = 4$
   $y = 4 - 1$
   $y = 3$

1. Rewrite $x + y = 4$ in terms of y ($y = 4 - x$).
2. Now substitute what y equals for the y variable of the second equation.
3. Now simplify by distributing the negative sign through the parentheses and solve for x.
4. Now substitute the $x = 1$ into the first equation to solve for y.

   Notice that the ordered pair is the same as above (1, 3). Substitution is just another way of solving for the point where these two equations intersect.
   See the graph on the preceding page.

Sometimes the elimination method of linear equations is not easy to use because the operations are same. See the example below:

1. $x + y = 13$
   $x + 2y = 1$
2. $x + y = 13$
   $-(x + 2y = 1)$

   $x + y = 13$
   $\underline{-x - 2y = -1}$
3. $-y = 12$
   $y = -12$
4. $x - 12 = 13$
   $\underline{+12+12}$
   $x = 25$

1. Here the operations in both equations are addition; therefore, eliminating a variable cannot be done the way that it is. However, the x's have the same coefficients.
2. So if we multiply one of the equations by -1, the x's would negate, making it easy to solve for the y variable.
3. Notice that the y is negative. In the final solution, it is conventional to change a negative variable to a positive variable. So again, multiply the equation through by -1 to make the y variable positive and the 12 negative.
4. Now substitute -12 into the first equation to find the value of x.

Sometimes the elimination method of linear equations is not easy to use because the coefficients of each equation are not the same. See the following example:

1. $5x - 2y = 4$
   $2x + 3y = 13$
2. $3(5x - 2y = 4)$
   $2(2x + 3y = 13)$
   ↓
3. $15x - 6y = 12$
   $\underline{4x + 6y = 26}$
   $19x = 38$
   $19x/19 = 38/19$
   $x = 2$
4. $5(2) - 2y = 4$
   $10 - 2y = 4$
   $\underline{-10 \quad -10}$
   $-2y = -6$
   $-2y/2 = -6/2$
   $y = 3$

1. Notice that the coefficients are not the same, so subtracting will not negate one of the variables.

2–3. In order to eliminate one of the variables, it is necessary to alter both equations. Multiplying the first equation by 3 and the second equation by 2 will allow you to negate the y values, leaving $19x = 38$, thus $x = 2$.

4. Now take the $x = 2$ and substitute it into the first equation to find the value for y.

Applications

Example: If one number is four times as large as another number, and the smaller number is decreased by two, the result is fourteen less than the larger number. What are the two numbers?

Answer: Let y be the smaller number and let x be the larger number.

$x = 4y$ Substituting 4y for x in the second equation we have:

$y - 2 = x - 14$
↓
$y - 2 = (4y) - 14$
$y - 2 = 4y - 14$
$\underline{-4y \quad -4y}$
$y - 4y - 2 = -14$
$\underline{\quad\quad +2 \quad +2}$
$-3y = -12$
$y = 4$

If $y = 4$, then $x = 4y$
$x = 4 \cdot 4$
$x = 16$

So the answers are: $x = 16, y = 4$.

Read the problem through to identify and assign variables. We are talking about two numbers, one smaller than other. Because we know little about these numbers, we use variables to represent them.

  y = the smaller number
  x = the larger number

Now read and translate the English phrases to algebraic expressions. Look for commas because commas usually indicate the completion of one operation and the beginning of another.

First, "If one number is four times as large as another number." This means a number has the same value as 4 times another number. This means I should have a

(continued)

Check: $y - 2 = x - 14$
$4 - 2 = 16 - 14$
$2 = 2$

proportion because proportions are equal to each other. So a larger value is equal to 4 times a smaller value. This seems reasonable. $x = 4y$.

Now for the next phrase, "the smaller number decreased by 2." We assigned the smaller value as y and decrease means to subtract by the number that is given, 2, so we have $y - 2$. Now we have one algebraic expression and one algebraic equation.

The last phrase says that "the result is fourteen less than the larger number." "Result is" means that the result of $y - 2$ equals 14 less than the larger number. So the algebraic expression is written like this: $y - 2 = x - 14$. We can substitute the x value into this equation and solve for y.

25

Roots, Powers, and Scientific Notation

INTRODUCTION

Just as multiplication is a more complex form of addition and division is a more complex form of subtraction, powers are a more complex form of multiplication and roots have the same relationship to division. That is, if you know and understand multiplication and division, then you will be able to understand powers and roots. The uses of both procedures are extensive. For example, in Chapter 18 (on the Pythagorean Theorem), each side is squared (raised to the power of two).

RELEVANT CONCEPTS FOR ALL TESTS

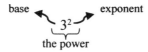

First, let's go through notation. To raise to a power means that the exponent tells the base how many times it should be written as a factor when multiplied by itself. For example:

$3^2 = 3 \cdot 3 = 9$ $\quad\quad$ $\begin{cases} not\ 3 \cdot 2 \\ not\ 4 \cdot 3 \end{cases}$
$4^3 = 4 \cdot 4 \cdot 4 = 64$

Now before roots can be properly explained, it is important that the components are identified. The radical is the symbol that is over the radicand. The index is the nth root that is being performed. For example:

$\sqrt[3]{27} = 3$ (because $3 \cdot 3 \cdot 3 = 27$)

with index, radical, radicand, and factors labeled.

Roots are the inverse operation of powers. The nth root is asking to find the factor that is multiplied by itself that will result in the radicand. For example:

$\sqrt{16} = \pm 4$ (because $4 \cdot 4 = 16$ and $-4 \cdot -4 = 16$)

Note: When *no* index is given, it is understood to be 2 (the square root).

$\sqrt{16}$ is equivalent to $\sqrt[2]{16}$

Here is an example using exponents:

Example: What is the value of $3x^2 - x + 1$ if $x = 4$? (Remember to follow the order of operations.)

Answer: $3x^2 - x + 1$
$3(4)^2 - 4 + 1$
$3(16) - 4 + 1$
$48 - 3$
45

Example: In the example above, what is the value if $x = -4$?

Answer: $3x^2 - x + 1$
$3(-4)^2 - (-4) + 1$
$3(16) + 4 + 1$
$48 + 5$
53

Example: If $x^2 - 1 = 80$, what is the value of x?

Answer: $x^2 - 1 = 80$
$x^2 - 1 + 1 = 80 + 1$
$x^2 = 81$
$\sqrt{x^2} = \sqrt{81}$
$x = \pm 9$

Notice that in both examples, the order of operations is *always* considered.
This is a good time to note that often there are *two* roots because in the above example, $-9 \cdot -9 = 81$, just as $9 \cdot 9 = 81$.

SCIENTIFIC NOTATION

Scientific notation is used as a simplified method to present very large or very small numbers by indicating a decimal times a positive power of 10 for large numbers or a negative power of 10 for small numbers.

18,000,000,000 is equivalent to 1.8×10^{10}

standard notation   scientific notation

The exponent indicates *the number of places* the decimal is moved, *not* the number of zeroes that are added to the value. In the example above, the decimal is moved ten places to the left (10^{10}). To write a standard notation into scientific notation, the decimal is placed behind the first digit, which results in the number being greater than or equal to 1 but less than 10 and then multiplied by the power of 10. Also:

.0000000033 is equivalent to 3.3×10^{-9}

standard notation   scientific notation

The decimal is moved 9 places to the right, hence the exponent of *negative* 9.

Example: Which is larger, 3,100,000 or 3.0×10^5?

Answer: The first number is larger because $3.0 \times 10^5 = 300,000$ (take 3.0 and move the decimal 5 places to the right to compare)

Example: 62.8 written in scientific notation would be:

Answer: 6.28×10^1 (6.28 moves the decimal one place to the left)

Example: 9.73×10^8 in standard notation is:

Answer: 973,000,000 (moving the decimal 8 places to the right)

Example: -2.7×10^5 is:

Answer: −270,000

Example: 863,000 expressed in scientific notation is:

Answer: 8.63×10^5

Example: 7.83×10^{-2} in standard notation is:

Answer: .0783

Example: 95,500,000 in scientific notation is:

Answer: 9.55×10^7

Example: 2^6 is:

Answer: $2^6 = 2 \cdot 2 \cdot 2 \cdot 2 \cdot 2 \cdot 2 = 64$

Example: $10^4 \div 10^2$ is:

Answer: $10^4 \div 10^2 = 10^{4-2} = 10^2 = 100$

Example: -6^2 is:

Answer: $-6^2 = -6 \cdot -6 = 36$ (not -36)

Example: $(-12)^2$ is:

Answer: $(-12)^2 = (-12)(-12) = 144$

Example: $-(12^2)$ is:

Answer: $-(12^2) = -(144) = -144$

Example: What is another way to write 8.75×10^3?

Answer: Move the decimal three places to the right: 8,750.

END OF PRAXIS I, 0014, 0511, AND 0146

Here are some further rules of exponents:

The easiest way to think of exponents is that each time an exponent is increased by one, the base is also increased by the value of the base, and every time the exponent is decreased by one, the base is decreased by the fractional part of the base (1/b). See the following pattern using 4 as a base:

Roots, Powers, and Scientific Notation

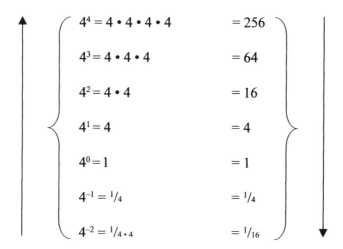

Each time the exponent increased by one, the product of the base is 4 times as great. Each time the exponent decreases by one, the product of the base decreases in value by $1/4$.

Rules for exponents in layman's terms:

1. Any base to the zero power always equals 1 ($a^0 = 1$).

2. Any base to the first power is equal to the value of the base ($a^1 = a$).

3. Any negative exponent is equal to the reciprocal of the base ($a^{-n} = 1/a^n$). Notice that the negative sign on the exponent is removed.

4. A root can be rewritten as a positive power with a fractional exponent where the index is the denominator and the exponent is the numerator ($a^{m/n} = \sqrt[n]{a^m}$).

5. When multiplying two powers having the same base, keep the base and add the exponents ($a^p \cdot a^q = a^{p+q}$). *Note the important distinction between adding and multiplying exponents.*

    When adding two powers, the variables and exponents must be the same before addition can be performed, and then only the coefficients are added.

    $8x^3 + 5x^3 = (8 + 5)x^3 = 13x^3$ ⬅——— adding
    $8x^3 \cdot 5x^3 = (8 \cdot 5)x^{3+3} = 40x^6$ ⬅——— multiplying

6. When a power is raised to a power, the exponents are *multiplied*.
   $(a^p)^q = a^{pq}$ $(7^2)^3 = 7^{2 \cdot 3} = 7^6$

7. When dividing exponents, *subtract* the exponents and keep the base.
   $a^p/a^q = a^{p-q}$

    Shortcut: Subtract the smaller-value exponent from the larger-value exponent, then put the difference where the subtraction takes place.

Example: What is the value of 2^{-4}?

Answer: $2^{-4} = 1/2^4 = 1/16$

Example: $4^{3/2}$ is:

Answer: $4^{3/2} = \sqrt[2]{4^3} = \sqrt{64} = 8$

Example: $2^3 \cdot 2^2$ is:

Answer: $2^3 \cdot 2^2 = 2^{3+2} = 2^5 = 64$

Example: $5^{-3} \cdot 5^7$ is:

Answer: $5^{-3} \cdot 5^7 = 5^{-3+7} = 5^4 = 625$

Example: $(2^4)^3$ is:

Answer: $(2^4)^3 = 2^{4 \cdot 3} = 2^{12} = 4,096$

There are corresponding rules that deal with radicals ($\sqrt{\ }$) or roots.

1. When the radicand expression is raised to a power that is the same as the index of the radical, it means to eliminate the radical and keep the value of the radicand.

   $\sqrt[n]{a^n} = a$

2. Only when the indexes are the same is factoring allowed.

   $\sqrt[n]{ab} = \sqrt[n]{a} \cdot \sqrt[n]{b}$

   By the same token, only when the indexes are the same can multiplication be performed.

   $\sqrt[n]{a} \cdot \sqrt[n]{b} = \sqrt[n]{ab}$

   Note: There are different rules for adding and subtracting radicals. Radicands and indexes have to be the same (e.g., $\sqrt[n]{b} + \sqrt[n]{b}$). This does not mean that the radicands are added or subtracted. It means the coefficients only. For example:

   $\sqrt{a} + \sqrt{a} = 2\sqrt{a}$ 　　In front of the radical is an understood 1 when there is no actual number written.

   1 is understood
   ↓
   $\sqrt{2} + 3\sqrt{2} = 4\sqrt{2}$ 　　Therefore, the coefficients are added *only*. Here
   ↓　　　　　　　　　　　　there are two radical expressions combined as one.
   $1\sqrt{2} + 3\sqrt{2} = 4\sqrt{2}$

Roots, Powers, and Scientific Notation

Do not confuse the product rule with combining like terms. *The root of a sum* does not equal *the sum of the roots*.

$\sqrt{9+16} \neq \sqrt{9} + \sqrt{16}$

because

$\sqrt{9+16} \neq \sqrt{9} + \sqrt{16}$

$\sqrt{25} \neq 3 + 4$

$5 \neq 7$

3. Two terms joined by an addition or subtraction sign cannot be separated into two radicals because when they are simplified, the solutions are not the same. The "roots of a sum" in the example above is 5 and "the sum of the roots" is 7. So this results in a false statement and is not allowed. This is different than multiplying roots (the product rule).

$\sqrt[n]{a/b}$

The above form says that the numerator "a" and the denominator "b" *are both* under the radical and cannot be reduced. It could be rewritten as:

$\dfrac{\sqrt[n]{a}}{\sqrt[n]{b}}$

However, it cannot stay in this form because a radical cannot be in the denominator in its *simplest form*. This means the denominator has to be *rationalized*.

How to rationalize the denominator of a radical expression after it has been rewritten:

$\dfrac{\sqrt{a}}{\sqrt{b}} \cdot \dfrac{\sqrt{b}}{\sqrt{b}}$ Multiply both numerator and denominator by the denominator.

$\dfrac{\sqrt{ab}}{(\sqrt{b})^2}$ The denominator becomes a square and the numerator becomes the product of a and b.

$\dfrac{\sqrt{ab}}{b}$ Now apply the first rule to get "b" out from under the radical. *The radical expression is now in its simplest form.*

4. Multiplying radicals with the *same* index and *different* powers follows the same rules as multiplying *two* powers. Instead of keeping the bases, the radicands are kept and the exponents added.

$\sqrt[n]{a^m} \cdot \sqrt[n]{a^n}$

$\sqrt[n]{a^{m+n}}$

$a^{(m+n)/n}$

$a^{m/n+1}$

Example: $2\sqrt{7}$ is:

Answer: Since $2^2 = 4$, then $2\sqrt{7} = \sqrt{4 \cdot 7} = \sqrt{28}$

Example: $25^{1/2}$ is:

Answer: $25^{1/2} = \sqrt[2]{25^1} = \sqrt{25} = 5$

Example: $15\sqrt{2} - 3\sqrt{2}$ is:

Answer: $15\sqrt{2} - 3\sqrt{2} = (15 - 3)\sqrt{2} = 12\sqrt{2}$

Example: $\sqrt{28}$ is:

Answer: $\sqrt{7 \cdot 4} = 2\sqrt{7}$

Example: $(4.3 \times 10^{-3})(4.7 \times 10^5)$ is:

Answer: $(4.3)(4.7)(10^{-3+5}) = 20.21(10^2) = 20.21(100) = 2{,}021$

Example: $\dfrac{6.4 \times 10^{-6}}{3.15 \times 10^{-3}}$ is:

Answer: $\dfrac{6.4 \times 10^{-6}}{3.15 \times 10^{-3}} = \dfrac{6.4}{3.15} \times 10^{-6-(-3)} = 2.03 \times 10^{-6+3} = 2.03 \times 10^{-3} = 2.03/10^3$

(Note: $10^{-3} = 1/10^3$)

Example: $\dfrac{\sqrt{20}}{\sqrt{35}}$ is:

Answer: $\dfrac{\sqrt{20}}{\sqrt{35}} = \sqrt{20/35} = \sqrt{4/7} = \dfrac{\sqrt{4}}{\sqrt{7}} = \dfrac{\sqrt{4}\sqrt{7}}{\sqrt{7}\sqrt{7}} = \dfrac{2\sqrt{7}}{7}$

26

Inequalities

INTRODUCTION

Older students will recall a time when math was performed in terms of equalities, and inequalities played no significant role. That is, students worked almost exclusively with statements about what quantities equaled what other quantities. Happily, nearly all the old rules that applied to equalities apply to inequalities.

RELEVANT CONCEPTS FOR ALL TESTS

First, let's get the new symbols clearly defined.

$>$ greater than $(9 > 5)$
$<$ less than $(12 < 15)$
$\geq$ greater than or equal to $(9 \geq x)$
$\leq$ less than or equal to $(5 \leq y)$

Two or more symbols may often be combined in one statement. Here is an example: $5 > x > 1$ (x is less than 5 and greater than 1).

The one important rule that makes inequalities different from equalities is: When both sides are multiplied or divided by a negative number, the inequality symbol reverses ($>$ becomes $<$, etc.).

Example: Solve for x. $5x - 6 > 14$

Answer: $5x - 6 + 6 > 14 + 6$
$5x > 20$
$5x/5 > 20/5$
$x > 4$

Notice that algebraically the manipulation is the same. You simply isolate the variable on one side of the inequality symbol instead of an equals sign, and then solve for the variable. This inequality *is not* divided by a negative, so the inequality symbol does *not* reverse. What does x > 4 represent? The solution tells us that any real number greater that 4 will solve this inequality.

See the number line below. It shows that any real number greater than 4 to infinity is a solution. Symbol notation is (4, ∞).

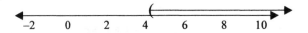

Notice that the array is a little beyond the 4 because the inequality symbol says that it does not include the 4.

Let's test numbers on either side of the 4 to verify that numbers greater than 4 make the inequality statement above true:

x = 3	x = 4	x = 4.1
5x – 6 > 14	5x – 6 > 14	5x – 6 > 14
5(3) – 6 > 14	5(4) – 6 > 14	5(4.1) – 6 > 14
15 – 6 > 14	20 – 6 > 14	20.5 – 6 > 14
9 ≯ 14	14 ≯ 14	14.5 > 14
False	False	True
3 is not in the solution set because 9 is not greater than 14.	4 is not in the solution set because 14 is not greater than 14; 14 is equal to 14 and the statement asks for solutions greater than 14.	4.1 is in the solution set because 14.5 is greater than 14. Notice that a solution set can be any real number very close to the right side of the value 4 as long as it is not 4.

Example: Solve for x.   –4x + 6 > 2x + 30

Answer:
$$-4x + 6 > 2x + 30$$
$$\underline{-2x \qquad\quad -2x}$$
$$-6x + 6 > \qquad 30$$
$$\underline{\quad -6 \qquad\qquad -6}$$
$$-6x \quad > \quad 24$$
$$-6x/-6 > 24/-6$$
$$x > -4$$
$$x < -4 \text{ (reversed)}$$

Notice that this time the inequality is being divided by a negative number. Therefore, to keep the inequality a true statement, the inequality symbol must be reversed.

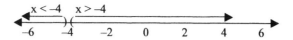

Let's test numbers on either side of −4 to verify why the x > −4 statement must reverse direction when dividing by a negative number.

$x > -4$		$x < -4$
Let x = −2 because −2 is greater than −4	Now	Let x = −5 because −5 is less than −4
−4x + 6 > 2x + 30		−4x + 6 > 2x + 30
−4(−2) + 6 > 2(−2) + 30		−4(−5) + 6 > 2(−5) + 30
8 + 6 > −4 + 30		20 + 6 > −10 + 30
14 > 26		26 > 20
False statement		True statement

This proves that when dividing by a negative, the inequality symbol must change direction in order to remain true. The same rule applies to multiplying by a negative number.

Example: Solve for x: $\dfrac{x}{2} - \dfrac{1}{3} > \dfrac{2x}{3} + \dfrac{1}{2}$ (Common denominator is 6)

$$\dfrac{\cancel{(6)}x}{\cancel{2}} - \dfrac{\cancel{(6)}1}{\cancel{3}} > \dfrac{\cancel{(6)}2x}{\cancel{3}} + \dfrac{\cancel{(6)}1}{\cancel{2}}$$

$$3x - 2 > 4x + 3$$

Answer: Instead of going through the process of changing all the denominators to 6 and then rewriting each fraction according to its new denominator, you can perform a shortcut as long as you do it to every term in the equation. Multiplying each term by the common denominator is known as "clearing the fraction." This allows you to get rid of the denominator for computational purposes (see above). Now finish solving the inequality.

3x − 2 > 4x + 3	Not all equations are solved vertically. This inequality is being solved horizontally. It's your prerogative.
3x − 4x − 2 > 4x + 3 − 4x	
−x − 2 > +3	
−x − 2 + 2 > +3 + 2	Also, notice that the inequality symbol has changed direction. Can you explain why?
−x > +5	
−x/−1 > 5/−1	
x < −5	

END OF PRAXIS I, 0014, AND 0146

Problems at this level are somewhat more complex but follow the same rules. The following is a type that might be found on PRAXIS tests 0069 or 0063, SAT, or ACT, where constructed response questions are featured.

Example: Let $a = 3$ and $b = 5$. Then if

$a < b$

$a(a) < a(b)$

$a^2 < ab$

$a^2 - b^2 < ab - b^2$

$(a + b)(a - b) < b(a - b)$

$(a + b) < b$

$(5 + 3) < 5$

Find the error and explain.

Answer: There is no error in the steps until both sides are divided by $(a - b)$. This quantity is negative and at that point, the symbol $<$ (less than) should have been replaced by $>$ (greater than).

1. You are given $a < b$, which means "a" is on the left of "b" on the number line; therefore, the value of "a" will be less than "b." So the number assignments are correct; $a = 3$ and $b = 5$.
2. Multiply both sides by "a," then simplify.
3. Subtract b^2 from both sides.
4. Both sides of the inequality symbol can be manipulated by factoring. On the left of the inequality equation is the difference of two squares. This gives $(a + b)(a - b)$. On the right of the inequality equation "b" can be factored out, making the expression $b(a - b)$. Both sides use legal operations.
5. Now arbitrarily substitute 3 for "a" and 5 for "b," then simplify.

1. $a < b$
2. $a(a) < a(b)$
   $a^2 < ab$
3. $a^2 - b^2 < ab - b^2$
4. $(a + b)(a - b) < b(a - b)$
5. $(3 + 5)(3 - 5) < 5(3 - 5)$
   $\dfrac{(8)(-2)}{(-2)} < \dfrac{5(-2)}{(-2)}$
   $(5 + 3) < 5$
   $\downarrow$
   $8 > 5$

Remember that when an inequality is divided or multiplied by a negative number, the inequality symbol changes direction.

Inequalities

There are several ways to represent inequalities using symbols and interval notation. It is important to be able to interpret the subtleness of a mathematical statement concerning inequalities.

Let a = some number
Let b = some number
$\}$ a < b "a" is always less than "b" on a number line

$-\infty$ represents negative infinity going left on a number line forever.

∞ represents positive infinity going right on a number line forever.

Parentheses stand for an open interval where the left parenthesis "(" indicates "greater than" (>) and the right parenthesis ")" indicates "less than" (<).

Brackets indicate a closed interval where the left bracket "[" indicates "greater than or equal to" (≥) and the right bracket "]" indicates "less than or equal to" (≤).

Symbolic notation is written as follows:

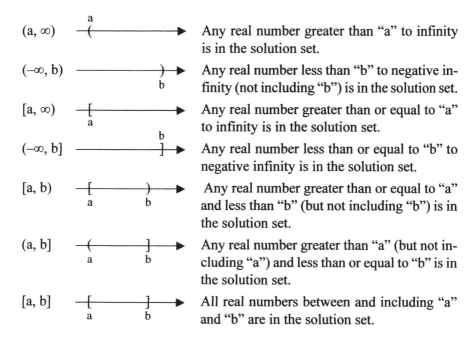

(a, ∞) Any real number greater than "a" to infinity is in the solution set.

(−∞, b) Any real number less than "b" to negative infinity (not including "b") is in the solution set.

[a, ∞) Any real number greater than or equal to "a" to infinity is in the solution set.

(−∞, b] Any real number less than or equal to "b" to negative infinity is in the solution set.

[a, b) Any real number greater than or equal to "a" and less than "b" (but not including "b") is in the solution set.

(a, b] Any real number greater than "a" (but not including "a") and less than or equal to "b" is in the solution set.

[a, b] All real numbers between and including "a" and "b" are in the solution set.

Absolute value inequalities follow the same steps as solving equations with absolute value; however, you must remember to reverse the direction of the inequality symbol when multiplying or dividing by a negative number, and when you rewrite the inequality to solve for the negative absolute value for x the inequality sign *must* be reversed as well, in order to keep the equation true. See the following example.

Example: |4x + 2| > 10

|4x + 2| > 10	or	|4x + 2| > 10
4x + 2 > 10		4x + 2 < –10
–2 –2		–2 –2
4x > 8		4x < –12
4x/4 > 8/4	or	4x/4 < –12/4
x > 2	or	x < –3

Check:

|4x + 2| > 10	|4x + 2| > 10
|4(3) + 2| > 10	|4(–4) + 2| > 10
|12 + 2| > 10	|–16 + 2| > 10
|14| > 10	|–14| > 10
14 > 10	14 > 10

Absolute value inequalities have a twist in how to solve for the x values. Notice that the absolute value equation indicates that the solution will be greater than 10 but does not include 10. This computation isn't new, but when the second absolute value inequality is rewritten to consider the negative value of x, not only is the negative sign used *but also the inequality sign is reversed*. Once this business is taken care of, solving for the x values is the same procedure as before using the property of equality. The solution set is (2, –3). This set represents the points on a number line where all values going in the direction of the arrays are solutions; however, the number line does *not* include the 2 or the –3 as part of the solution set, and the values between –3 and 2 are not solutions; thus the interval notation is (x < –3), (x > 2).

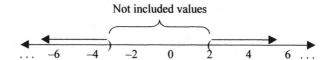

Remember that when checking inequalities, it is important to do what the inequality symbol says. Here it says the values to check have to be either greater than 2 (not to include 2) and less than –3 (not to include negative 3). This is an important idea because not only will the values less than –3 keep the statement true, but also values greater than 2 will keep the statement true. For example, values greater than 2 could be values like 2.01, and values less than –3 could be values like –3.01. Each value is slightly larger or smaller than the given x values.

Example: $|-4x + 6| \geq |2x + 30|$

$$|-4x + 6| \geq |2x + 30| \qquad\qquad |-4x + 6| \geq |2x + 30|$$

$$\begin{array}{l} -4x + 6 \geq 2x + 30 \\ \underline{-2x \qquad -2x} \\ -6x + 6 \geq \qquad 30 \\ \underline{\quad -6 \qquad -6} \\ -6x/-6 \geq 24/-6 \\ x \geq -4 \\ \downarrow \\ x \leq -4 \text{ (reversed)} \end{array} \qquad \begin{array}{l} -4x + 6 \leq -2x - 30 \\ \underline{+2x \qquad +2x} \\ -2x + 6 \leq \qquad -30 \\ \underline{\quad -6 \qquad -6} \\ -2x/-2 < -36/-2 \\ x \leq 18 \\ \downarrow \\ x \geq 18 \text{ (reversed)} \end{array}$$

```
         x ≤ -4                                              x ≥ 18
  ◄──────────────┤                                    ├──────────────►
      -6   -4   -2    0   ......   12   14   16   18   20
```

Notice that we had to utilize both rules in this example. First, we divided the inequality by a negative, so in order to keep the inequality statement true, the inequality symbol had to be reversed from $x \geq -4$ to $x \leq -4$. Second, when solving absolute value equations (and thus inequalities), both negative and positive x values must be considered. Therefore, the absolute value inequality here had to be rewritten as an inequality without the absolute value bars to consider the negative x value. Because we changed the positive sign of the rewritten inequality to a negative sign, we also had to change the direction of the inequality symbol from $\geq$ to $\leq$. The number line displays the solution set for each. All values from positive 18 to positive infinity $[18, \infty)$ are possible solutions and all values from negative infinity to -4 $(-\infty, -4]$ are also possible solutions.

Check:

$\lvert -4x + 6 \rvert \geq \lvert 2x + 30 \rvert$	$\lvert -4x + 6 \rvert \geq \lvert 2x + 30 \rvert$
$\lvert -4(-4) + 6 \rvert \geq \lvert 2(-4) + 30 \rvert$	$\lvert -4(18) + 6 \rvert \geq \lvert 2(18) + 30 \rvert$
$\lvert 16 + 6 \rvert \geq \lvert -8 + 30 \rvert$	$\lvert -72 + 6 \rvert \geq \lvert 36 + 30 \rvert$
$\lvert 22 \rvert \geq \lvert 22 \rvert$	$\lvert -66 \rvert \geq \lvert 66 \rvert$
$22 \geq 22$	$66 \geq 66$

Notice that using the exact values (-4 from the solution $x \leq -4$ and 18 from the solution $x \geq 18$) is perfectly acceptable because the inequality symbol says that for the solution $x \geq 18$ it will accept values that are *equal to or greater than* 18, and it also says for the solution $x \leq -4$ it will accept values that are *equal to or less than* -4. Both checks make true statements.

Special Cases of Absolute Value Inequalities: True or False

$|x| > -1$ Is this a true statement?

Yes, it is. Remember that with absolute value, an x value could be negative or positive. We also know that whenever the absolute value is taken of some number, the solution will always be greater than or equal to zero, so if the absolute value is = 0, then $0 > -1$ is a true statement.

$|x| < -1$ Is this a true statement?

No, it is not, because the "less than" symbol says that the absolute value of some number will be less than zero. This *cannot* happen. Remember that the absolute value of any number will always be positive or zero. Therefore, any value of x, no matter how large or small, will give an absolute value of zero or greater but never less than -1, as in the statement above.

$-|x| < 1$ Is this a true statement?

Yes, it is. After getting the absolute value of any number (x), multiplying the x value by a negative will give a negative product. For example:

When x = 2   When x = 0
  $-|2| < 1$ $-|0| < 1$
   ↓ ↓   ↓ ↓
$(-1)(2) < 1$ $(-1)(0) < 1$
$-2 < 1$ $0 < 1$ (remember, 0 has no sign)

When an inequality is greater than (>) or greater than or equal to ($\geq$), it is clear whether or not the statement is true. The problem arises when the inequality is less than (<) or less than or equal to ($\leq$). You have to understand what you are asked!

27

Age Problems

INTRODUCTION

Age problems have almost achieved status as icons of difficult math. Jokes abound centered around problems that begin "If Jane is twice Joe's age and fifteen years ago . . ." You are sure to have heard many yourself. Let's see how to approach them so they will present no particular challenge to you on a standardized test.

RELEVANT CONCEPTS FOR ALL TESTS

Age problems have always given students fits. The key is to summarize the data in a table, derive a formula, and then solve. Make sure that you have an approximate answer clearly in mind. Checking is also not very hard and is strongly recommended.

Example: Thad and Paul are brothers. Paul is thirty-five years old. Three years ago, Paul was four times as old as Thad was then. How old is Thad now?

Answer: First, make a table that summarizes the data.

	Now	3 years ago
Paul	35	(35 − 3) = 32
Thad	t	t − 3

Next, write the formula that states the relationship three years ago.

$32 = 4(t - 3)$
$32 = 4t - 12$
$32 + 12 = 4t$
$44 = 4t$

44/4 = t
11 = t

Thad is 11 years old now and was 8 three years ago.

Let's check our answer. Is 8 one-fourth of 32? Yes, so the answer is correct. Remember, it is always good to make a quick check of the answer.

Example: Kay is three times as old as Samantha. The sum of their ages is 24. How old is Kay?

Answer: First, make the table.

	Now
Kay	3x
Samantha	x

Next, write the formula.

3x + x = 24
4x = 24
x = 24/4
x = 6

Samantha is six years old, Kay is 3 × 6, or 18.

Let's check our answer. If Samantha is 6 and Kay is 18, the sum of 6 and 18 is 24. Our answer checks and should be correct.

Example: Matthew is six years older than Herman. In two years, Matthew will be twice as old as Herman. How old is Herman now?

Answer: First, make the table.

	Now	**In two years**
Matthew	x + 6	x + 8
Herman	x	x + 2

Next, write the formula.

x + 8 = (x + 2)2
x + 8 = 2x + 4
8 − 4 = 2x − x
4 = x

Herman is 4 years old.

Let's check our answer. If Herman is 4, Matthew is 10. In two years, Herman will be 6 and Matthew 12. 6 × 2 = 12. The answer checks.

Example: Betty's age is three times Marion's. If you were to add 20 to Marion's age and subtract 20 from Betty's, the ages would be the same. How old are they today?

Answer: First, make the table.

Betty   3x
Marion   x

Next, write the formula.

$x + 20 = 3x - 20$
$x - 3x = -20 - 20$
$-2x = -40$
$2x = 40$
$x = 40/2 = 20$

That makes Marion 20 years old and Betty 60.
   Checking to be sure, $60 - 20 = 20 + 20$
                          $40 = 40$

The answer is correct.

Example: A French statue is three times as old as a Canadian statue. In 100 years the French statue will be only twice as old as the Canadian one. How old are they now?

Answer: First, make the table.

	Now	In 100 years
French statue	3x	3x + 100
Canadian statue	x	x + 100

Next, write the formula.

$2(x + 100) = 3x + 100$
$2x + 200 = 3x + 100$
$2x - 3x = 100 - 200$
$-x = -100$
$x = 100$

So the French statue is 300 years old and the Canadian one is 100. In 100 years the French statue will be 400 years old and the Canadian one 200. That checks because 200 is half of 400.

Example: George is 22 years older than Mike. When Mike is as old as George is now, he will be three times his present age. How old are George and Mike now?

Answer: First, make the table.

	Now	**In 22 years**
George	x + 22	x + 44
Mike	x	x + 22

Next, write the formula.

3x = x + 22
3x − x = 22
2x = 22
x = 11

George is 33 years old and Mike is 11.
   Do these numbers check? 11 + 22 = 33
                          33 = 33   Yes, the answer checks.

Example: Maggie is shy about revealing her age but says that her sister Katherine, who is three years older, was twice her age ninety years ago. How old are Maggie and Katherine now?

Answer: First, make the table.

	Now	**90 years ago**
Maggie	x − 3	(x − 3) − 90
Katherine	x	x − 90

Next, write the formula.

(x − 90)/2 = (x − 3) − 90
x − 90 = 2x − 6 − 180
−90 + 6 + 180 = 2x − x
96 = x

So Katherine is 96 years old and Maggie is 93.
   Does it check? 90 years ago the girls would have been 3 and 6, so Katherine would have been twice as old. It does check.

Example: Jerome is half his calculus teacher's age. In 20 years he will be two-thirds his teacher's age. How old are they now?

Answer: First, make the table.

	Now	**In 20 years**
Jerome	x	x + 20
His teacher	2x	2x + 20

Next, write the formula.

$(2x + 20)2/3 = x + 20$
$(2x + 20)2 = 3x + 60$ (multiply both sides by 3)
$4x + 40 = 3x + 60$
$4x - 3x = 60 - 40$
$x = 20$

Jerome is 20 years old and his teacher is 40.

Let's confirm the answer. In twenty years Jerome will be 40 and his teacher 60. 40 is two-thirds of 60. The answer is correct.

END OF PRAXIS I, 0014, 0511, AND 0146

These examples are a bit more complex but the procedures are very similar to the preceding ones.

Example: Jill is 12 years older than her sister Rose, and the product of their ages is 540. How old are they?

Answer: First, make the table.

Jill	$x + 12$
Rose	x

Next, write the formula.

$x(x + 12) = 540$
$x^2 + 12x = 540$
$x^2 + 12x - 540 = 0$
$(x + 30)(x - 18) = 0$
$x + 30 = 0 \quad x - 18 = 0$
$x = -30 \quad x = 18$

The negative answer can be discarded on practical grounds, so Rose is 18 years old and Jill is 30.

Let's confirm the answer. Does $18(18 + 12) = 540$?
$$18(30) = 540$$
$$540 = 540 \text{ The answer is correct.}$$

28

Trigonometry

Note: This content is *not* related to PRAXIS I, 0014, 0511, or 0146.

INTRODUCTION

The right triangle is the basis for trigonometry. Because the two acute angles are complementary (that is, their sum equals 90°), all the sides of a right triangle may be determined from a small amount of information. For example, one acute angle and one side will determine the other sides and the third angle.

Here is the basic terminology used in trigonometry:

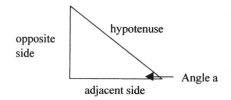

Here are the basic side length relationships:

Sine = opposite/hypotenuse

Cosine = adjacent/hypotenuse

Tangent = opposite/adjacent

If you find mnemonics helpful, here is one to help with the above:

SOH CAH TOA

Trigonometry

Think: **S**ine = **o**pposite/**h**ypotenuse = SOH
Cosine = **a**djacent/**h**ypotenuse = CAH
Tangent = **o**pposite/**a**djacent = TOA

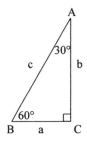

Angle A is opposite the short leg (a)
Angle B is opposite the long leg (b)
Angle C is opposite the hypotenuse (c)

Trigonometry may use 30°–60°–90° and 45°–45°–90° right triangles. Both of these triangles will help you to solve for the side lengths of the most common trigonometric ratios.

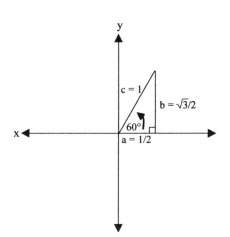

This 30°–60° triangle is based on a unit circle (see below), meaning the radius will always be 1. The radius is the hypotenuse. With this triangle we can find the six basic trigonometric ratios without using the Pythagorean Theorem or even resorting to the calculator.

First, let's find the sine, cosine, and tangent of 60°.

Sin 60°	$\frac{opp}{hyp}$	Start at angle 60°. Look opposite to "b," then look for the hypotenuse "c."	$\frac{\sqrt{3}/2}{1}$	Sin 60° = $\sqrt{3}/2$	Any number divided by 1 is unchanged.
Cos 60°	$\frac{adj}{hyp}$	Start at angle 60°. Look opposite to "a," then look for the hypotenuse "c."	$\frac{1/2}{1}$	Cos 60° = 1/2	Any number divided by 1 is unchanged.
Tan 60°	$\frac{opp}{adj}$	Start at angle 60°. Look opposite to "b," then look adjacent to "a."	$\frac{\sqrt{3}/2}{1/2}$	Tan 60° = $\sqrt{3}$	

The tan function turns out to be a complex fraction. Because the denominators are the same, the shortcut is to cancel the 2's. This simplifies to √3/1 or simply √3.

Next, let's find the sine, cosine, and tangent of 30°. This takes the same idea and adds a few modifications. Instead of having "b" on the y axis it will be on the x axis and "a" will no longer be on the x axis; instead it will be on the y axis. Cosine and sine reverse with the hypotenuse remaining 1. Remember, the hypotenuse is the radius on a unit circle.

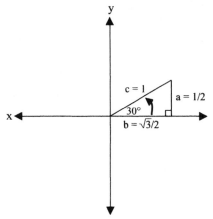

Following the same process from above you will see that

$\sin 30° = 1/2$

$\cos 30° = \dfrac{\sqrt{3}/2}{1} = \dfrac{\sqrt{3}}{2}$

$\tan 30° = (1/2)/(\sqrt{3}/2) = \sqrt{3}/3$

Now let's see the behavior of a 45°–45°–90° triangle. You follow the same procedure, except the acute angles will be 45°. Remember, a triangle with two equal angles is an isosceles triangle. In this case it is a right isosceles triangle. This means that the sides (legs) will have the same side lengths; therefore, either leg divided by the hypotenuse (sine and cosine) will render the same value. So the legs divided by each other (tangent) will always be equal to 1.

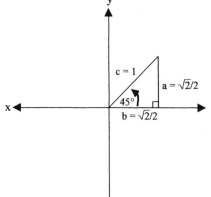

Following the same process from above, you will see that:

$\sin 45° = \dfrac{\sqrt{2}/2}{1} = \sqrt{2}/2$

$\cos 45° = \dfrac{\sqrt{2}/2}{1} = \sqrt{2}/2$

$\tan 45° = \dfrac{2}{\sqrt{2}/2} = 1$

The sine, cosine, and tangent have reciprocal functions.

Trigonometry

Sine	opp/hyp	→ Cosecant	hyp/opp	A particular function has a reciprocal. For example:
Cosine	adj/hyp	→ Secant	hyp/adj	Sin 30° = 1/2 Then Cosecant = 2/1 → 2
Tangent	opp/adj	→ Cotangent	adj/opp	Cos 30° = √3/2 Then Secant = 2/√3 → $\frac{2(\sqrt{3})}{3}$

The denominator needed to be rationalized (this removed the radical from the denominator).

Example: Given the following right triangle, what are the other side lengths?

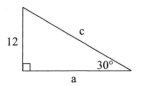

Answer: Step 1: Solving for hypotenuse (c) using sin 30° (because opp/hyp will isolate the variable "c."

Sin 30° = ½ = .5

Sin 30° = 12/c

c • sin 30° = 12 • ̸c / ̸c

c • sin 30° = 12

c • sin 30̸° / s̸in 30° = 12 / sin 30°

c = 12 / ½

(12 • 2 = 24)

c = 24

1. Divide both sides by "c," then cancel the c's on the right to get 12 and multiply "c" by sin 30° on the left.

2. Divide both sides by sin 30°, then cancel the sin 30° on the left to leave "c" and divide 12 by sin 30° on the right to get 12. Then divide 12 by ½, which is sin 30°.

3. Multiply 2 by 12 (the short leg) to get the length of the hypotenuse. (Remember, dividing by ½ is the same as multiplying by 2.)

Note: all you had to do is use the 30°–60° triangle to find sin 30° without using the calculator. If you choose to use the calculator, just enter sin (30°) then press enter. It will give the decimal value. (Make sure you use parentheses on both sides of the 30°.)

Step 2: Solve for the unknown leg (a) using Cosine 30° (because adj/hyp is the best ratio that will isolate the variable "a").

Cos 30° = √3/2

Cos 30° = a/24

√3/2 = a/24

Above is a step-by-step procedure to solve for a trigonometric function. However, this next example cuts many steps because if you know the value of cos 30°, you can simply use that ratio and solve for the variable "a."

$\sqrt{3}/2 \cdot 24 = a$

$12\sqrt{3} = a$ (or 20.78)

Of course, you can always use the Pythagorean Theorem if you know at least two sides of a right triangle.

Alternative Step 2 (Pythagorean Theorem)

$24^2 = 12^2 + a^2$

$576 = 144 + a^2$

$576 - 144 = a^2$

$432 = a^2$

$\sqrt{144 \cdot 3} = a$

$12\sqrt{3} = a$ (same as above and would serve as a good check)

Example: What is the area of the following triangle? (*Note*: it is not a right triangle.)

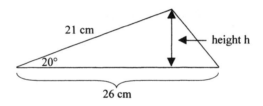

Answer: Step 1: find the height using the right triangle formed by the height with the hypotenuse = 21.

Sin 20 = .3420 = h/21

21(.3420) = h

7.18 cm = h

Step 2: Calculate the area.

A = bh/2

A = (26 • 7.18)/2

A = 13 • 7.18 = 93.37 cm²

EXPANSION TO THE UNIT CIRCLE

A limit on trigonometry based on the right triangle is that acute angles must be less than 90°. However, this figure is only one quarter of the total circle. When trigonometric functions are expanded to the full 360° and beyond, many other applications, such as electronics, are possible.

Here is how the expansion is performed. A circle is constructed with radius = 1 by the formula $x^2 + y^2 = 1$. Here is a rendition with quadrants marked.

Trigonometry

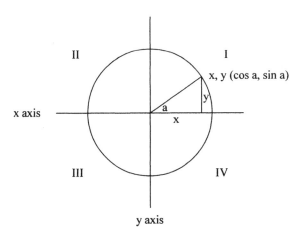

Inscribed in the circle could be an infinite number or right triangles like the one above, with one vertex at the center and a second on the circle.

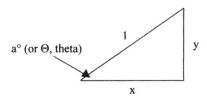

Each of these triangles can be described as having

$\cos a = x/1 = x$

$\sin a = y/1 = y$

$\tan a = y/x = \sin a/\cos a$

Most important, this point can be moved around the complete circle to find the sine and cosine, which are the x and y coordinates. Now such functions as the sine of 195° can be found, even though the ability to picture them as components of a right triangle is not possible.

Here are the signs for the functions by quadrants.

Quadrant	Cosine	Sine	Tangent
I	+	+	+
II	−	+	−
III	−	−	+
IV	+	−	−

Each point where an inscribed triangle touches the circle can be named as (x, y), but this will also be the same location as (cos a, sin a) because the radius (hypotenuse) is always 1.

The unit circle even allows the calculation of trigonometric functions of angles beyond 90°. This is true even though it is impossible to actually draw a triangle containing angles of that size. For example, it is possible to calculate the trigonometric values associated with an "acute" angle of 120. This is one of the amazing ways to use the unit circle and why it is so important to understand (sine 120° ≈ .866).

Even functions of angles exceeding 360° can be determined, even though they wildly exceed our ability to include them in any triangle!

Example: sine 405° = sine (405° – 360°) = sine 45° = $\sqrt{2}/2$ ≈ .707

Important Relationships

Angle	Sin	Cos	Tan
0°	0	1	0
30°	1/2	$\sqrt{3}/2$	$\sqrt{3}/3$
45°	$\sqrt{2}/2$	$\sqrt{2}/2$	1
60°	$\sqrt{3}/2$	1/2	$\sqrt{3}$
90°	1	0	U
120°	$\sqrt{3}/2$	–1/2	–$\sqrt{3}$
150°	1/2	–$\sqrt{3}/2$	–$\sqrt{3}/3$
180°	0	–1	0
270°	–1	0	U
360°	0	1	0

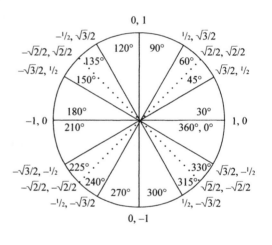

"U" denotes a function that is undefined because its calculation would necessitate division by zero, which is never allowed.

Important Identities

cos (a + 360) = cos a

sin (a + 360) = sin a

cos (–a) = cos a

sin (–a) = –sin a

$\sin^2 a + \cos^2 a = 1$

cos (90 – a) = sin a

sin (90 – a) = cos a

Note: Angle "–a" below simply denotes a clockwise rotation instead of the usual counterclockwise one.

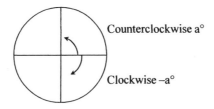

RADIANS

Angles may be expressed in degrees or radians, which are circle lengths in terms of radii.

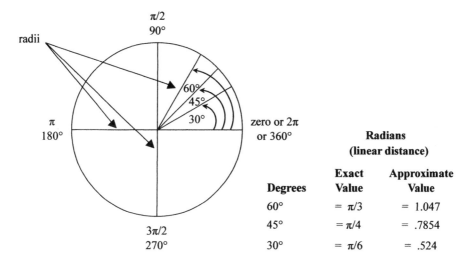

1 radian ≈ 57.3° (180/π)

Examples: Degrees to radians
60° = 60π/180 = 6π/18 = π/3 ≈ 1.047
30° = 30π/180 = 3π/18 = π/6 ≈ .524
45° = 45π/180 = π/4 ≈ .785
90° = 90π/180 = 9π/18 = 1π/2 = π/2 ≈ 1.57
120° = 2π/3 ≈ 2.094
180° = π ≈ 3.142

Remember how complex fractions work. For example:

$$\frac{60}{180/\pi} = \frac{60/1}{180/\pi} = \frac{60}{1} \cdot \frac{\pi}{180} = \frac{60\pi}{180}$$

29

Measurement

INTRODUCTION

It is likely that most standardized tests will have questions that include measurement, either in the customary system or in the metric system. Chapters 15–17 have some examples, but this chapter will bring together the various applications.

Often a question will simply involve the conversion of one measure to another. The following are equivalent in the customary system, which gradually evolved over centuries from ancient units.

Length	12 inches = 1 foot	
	36 inches = 1 yard	
	3 feet = 1 yard	
	5,280 feet = 1 mile	
	1,760 yards = 1 mile	
Volume	8 ounces = 1 cup	
	16 ounces = 1 pint	
	2 pints = 1 quart	
	4 quarts = 1 gallon	
Weight	16 ounces = 1 pound	
	2,000 pounds = 1 ton	

The metric system, which originated in France in 1791, is much more logical in its relationships. The basic units are meter (length), gram (weight), and liter (volume). These basic units all have the same prefixes, which denote the same value.

Prefix	Value
Milli-	1/1,000
Centi-	1/100
Deci-	1/10
Unit (meter, gram, or liter)	
Deca-	10
Hecto-	100
Kilo-	1,000

So, within the metric system conversions involve only the moving of the decimal. For example: 1 meter = 10 decimeters = 100 centimeters = 1,000 millimeters.

IMPORTANT CONSIDERATIONS

1. When calculating answers, make sure that you are using the correct units. For example, in a perimeter or circumference question, the answer is always in units (feet, centimeters, yards, miles, etc.), never square units or cubic units.
2. When calculating area, the units are always square units (square inches, inches^2, square meters, etc.), not units or cubic units.
3. When calculating volume, the units are always in cubic units (inches^3, yards^3, cubic yards, etc.), not units or square units.
4. Some conversions can lead to errors.
    a. 1 square foot = 144 square inches (not 12)
    b. 1 square yard = 9 square feet (not 3)
    c. 1 cubic yard = 27 cubic feet (not 3 or 9)

About the Authors

CHARLES W. HATCH, Ph.D., is President of CWH Consulting Company, Newberry, South Carolina. He earned a Master of Arts in Teaching at Johns Hopkins University and a doctorate in Educational Research and Measurement at the University of South Carolina. He has taught college courses in tests and measurement, statistics, and test preparation. Dr. Hatch has published the "Introductory Handbook of Measurement," "An Introductory Handbook for Statistical Package Programming," and papers on the subject of predicting freshman retention. He has served as a consultant on test preparation, college retention, and microcomputers and software.

Dr. Hatch is affiliated with the Friedman Institute for Evidence-Based Decision-Making in Education (EDIE). He authored *Pass That Test: A Guide to Successful Test Taking* (2008) and coauthored the *Educators' Handbook on Effective Testing* (2003) with Dr. Myles Friedman et al.

Dr. Hatch lives in Newberry, South Carolina, near his two children, Lisa and David, and his four granddaughters, Samantha, Kimberly, Robin, and Brooke.

MICKI DURHAM GIBSON, MAT, is a math educator with an undergraduate degree from Southern Wesleyan University and an MAT from Clemson University. She is currently a Ph.D. candidate in Educational Research at the University of South Carolina. She has conducted test preparation workshops across the Southeast from North Carolina to Mississippi. Her current interests include Singapore math, slide rules, statistics, test preparation, and teacher training.

Mrs. Gibson resides in Pendleton, South Carolina, with her husband, Steve, and her four children, Alison, Jason, Katherine, and John-Michael.